Creative Activities in Mathematics

PROBLEM-BASED MATHS
INVESTIGATIONS FOR LOWER
AND MIDDLE SECONDARY

BOOK 3

BY DEREK HOLTON AND CHARLES LOVITT

Published in 2025 by Amba Press, Melbourne, Australia
www.ambapress.com.au

First published in 2015 by ACER Press, an imprint of
Australian Council for Educational Research Ltd

© 2025 Derek Holton and Charles Lovitt

All rights reserved. No part of this book may be reproduced or transmitted in any form or by any means, electronic or mechanical, including photocopying, recording or by any information storage and retrieval system, without prior permission in writing from the publisher.

Cover design: Tess McCabe
Editor: Diane Fowler
Typesetting: Peter Long

ISBN: 9781923569263 (pbk)
ISBN: 9781923569270 (ebk)

A catalogue record for this book is available from the National Library of Australia.

TABLE OF CONTENTS

Introduction: The script and the performance	5
Part 1: Number and Algebra	**9**
Chapter 1: Number puzzles	10
Chapter 2: The Tower of Hanoi	28
Chapter 3: Nim-like games	43
Part 2: Measurement and Geometry	**57**
Chapter 4: Hidden treasure	58
Chapter 5: Tessellations	74
Chapter 6: How high is a building?	92
Part 3: Statistics and Probability	**103**
Chapter 7: Greedy Pig	104
Chapter 8: Pascal's triangle	115
Chapter 9: Monty Hall's problem	127

INTRODUCTION: THE SCRIPT AND THE PERFORMANCE

Who is this book for?

This book aims to provide extended activities that either introduce or prepare the way for material from the secondary section of the Australian Curriculum: Mathematics, or that consolidate material that has already been taught. All of the activities provide valuable opportunities for developing students' abilities within the curriculum's proficiency strands. It is for the use of:

- experienced teachers who work regularly in the area of mathematics
- teachers who may not be as confident in maths as in other areas
- relieving teachers needing additional maths support and class resources.

This is the third of three books in the *Creative Activities in Mathematics* series that cover most of the school years. Although Book 2 is largely for upper primary students, some of the material of that book can be used for secondary students too. Book 1 is for lower primary school students, but some of the activities there have been extended in this book. It is certainly possible to do the same for other Book 1 activities.

Notes for the director

All the world's a stage—but stages needs plays, and plays need scripts. This book presents several 'scripts'—creative maths activities—on the assumption that your classroom is as good a place to be a stage as any other. We want to tell a story of how problem solving might be used to foster interest in maths, to show that maths is more than learning new tricks, and to demonstrate how all of this fits into the Australian Curriculum.

But even Shakespeare's scripts get messed around from time to time. They may have been changed by the time they first hit the stage in Elizabeth's reign, and they have certainly been converted to modern dress, 'empty' sets, lavish films and musicals since then. So we don't expect our mathematical works to be used in precisely the form presented here. We expect them to be used in whatever way you think best for the class that you are currently with—and we expect them to evolve over time as a result of your experience with the 'lines' and where they lead you.

Why give a problem setting for the scripts? There are a number of reasons. First, mathematics is more than content to learn by heart. There has always been a creative element to the overall mathematical play. If this were not the case, how would we have ever found out about multiplication or decimals? Plays are more than scripts, and maths is more than content.

Second, problem solving helps develop reasoning and higher-order thinking in a way that pure repetition of exercises cannot do. This wouldn't matter much if there were no other plays than those enacted in school. But all the world's a stage, and on that stage no script is ever repeated. New situations arise and new problems have to be solved that were never dreamt of in school or even in university. This puts a premium on problem solving, reasoning and higher-order thinking. It is much easier to learn the basis of these skills (in all subjects) in school than in the drama of everyday life.

And third, there is a chance that banging their heads against new challenges in novel situations may be more interesting for students. It might even be more enjoyable and this extra motivation may increase learning and success.

The scripts

The plays here continue from those in *Creative Activities in Mathematics* Books 1 and 2. It may be worthwhile for some of your students to tackle the challenges in Book 2, and we have extended the odd problem from Book 1. As in the previous two books, the activities here aim to encourage students to see how mathematicians work and mathematics develops, by going through certain acts as the plays develop. Hopefully they will be encouraged to know that they can go through the same steps, listed in Figure 1.

Figure 1: Attacking problems

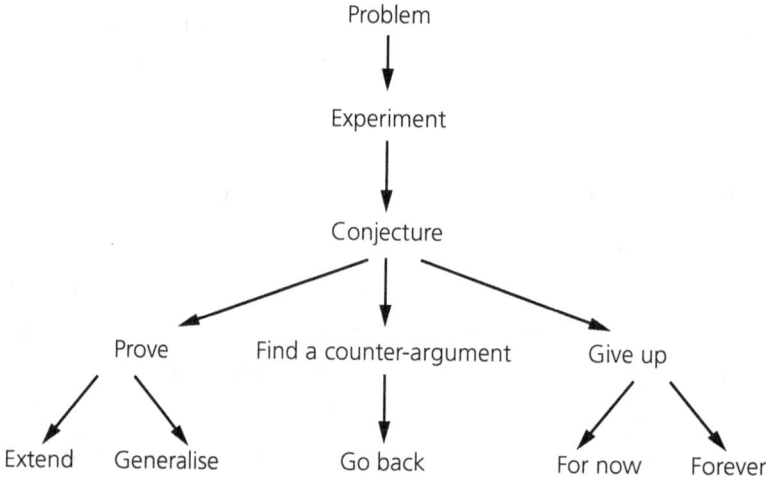

Every play has a basic *problem* that sets it on its course. Hopefully they are genuine problems in that students haven't seen them before and don't immediately know how to solve them. They may feel temporary panic because they can't see how they can ever solve this problem. They need to try to understand the problem and make sure they know what it is asking. From here they should do some guessing, checking and generally playing around with the problem to see what they can make of it. We think of this as *experimenting*. That process might lead to a reassessment of the problem and a feeling that they might be able to do something with it after all.

Although Figure 1 has definite arrows that purport to show the directions by which problems can be solved, it is hardly ever that easy. After experimenting, some problems can be solved right away, with no need for a conjecture and hardly any for a proof. In those cases we lead the students to look at other relevant situations that *extend* the problem in some way. For example, the original problem might use the number 3 in some significant way; one possible extension will involve using 4; later ones use 5 and so on. If we can get to a general answer for all numbers, that gives a *generalisation* that involves the original problem and presents the number 3 as a special or specific case.

But in many of the activities, students look at what might possibly happen. This is a *conjecture*. That word only means a 'guess'. Perhaps surprisingly, mathematicians guess a lot. This helps them make progress—it focuses their mind and helps them set up a plan of attack on the problem.

If your students are very lucky the conjecture can be proved and that proof may lead on to further work. If they are unlucky, two things could happen. Perhaps the best form of bad luck is that they might see something that goes against the conjecture and shows it can't be true—the dreaded *counter-example* or *counter-argument*. One way forward is to do more experimenting in the hope of finding a better conjecture. But if students can't make any progress, they may be forced to give up.

At this point the situation is in your hands. You may decide that they need some scaffolding to help them over the difficulty. Don't tell them what to do, though; they need to learn how to scaffold themselves. On the other hand you may decide to let them *give up for now*. A very useful problem-solving technique is to leave the problem alone for a while, until the idea of what to do next pops up and they can move on. You might, however, feel that the students have gone as far as they can go, and allow them to *give up forever* on the problem. Let them do this with honour, knowing that they have been successful on parts of the problem.

The layout of the scripts

As in the previous books in this series, this book has nine folios (chapters) that each contain three plays (activities) that are largely linked to one of the Australian Curriculum's content strands.

Each play is divided into four acts that we call 'levels'. These levels will be accessible to more than one curriculum year level, but as the play goes through the acts the year levels increase in difficulty slightly. This enables you to provide material for students across the secondary range, as well as giving the opportunity for one play to be used in one class with students entering and leaving as appropriate to their ability. Each activity develops through the levels in a series of scenes or 'steps' that move through the curriculum content descriptions and/or require more sophisticated work from students.

The Australian Curriculum: Mathematics is central to these activities. Each activity includes a table that lists relevant content descriptions. All of the activities here have a strong link to the Problem Solving proficiency strand. Some activities also address multiple content strands, the general capabilities and/or the cross-curriculum priorities of the Australian Curriculum.

Activity layout

The layout of each play or activity is the same. After giving the (starting) problem, we provide material under the following headings.

- Initial problem
- Background information
- Big ideas
- Suggested resources (if any)
- Problem aims
- Key concepts
- Possible heuristics/strategies
- Special notes
- Levels and steps

The *Initial problem* is the start of the activity for the students and gives some idea of the topic of the activity. This is followed by *Background information*, which ranges over interesting details about the problem, sometimes with an historical note, and how it might be taught. This section aims to give teachers an overview of the activity and how it develops from the initial problem. It sometimes notes any links with similar problems and related ideas. A table sets out the Australian Curriculum: Mathematics content descriptions for each level of each activity. (Note that only part of some descriptions may be covered by a given activity.) This table also sets out the way the problem develops and makes comments on the mathematics that the levels contain. At the end of this section we list the *Big ideas* of the activity.

It is probably self-evident what *Suggested resources*, *Problem aims*, *Key concepts* and *Possible heuristics/strategies* are. *Special notes* are used to provide a key definition or idea that is needed in the activity.

The activity is then broken up into its levels, which develop the problem based around questions and answers. Changes of directions or extensions of the problem are indicated by 'steps'. Throughout each problem there are questions that teachers might ask of students in the course of developing the activities.

Each level problem is completed by a section called 'Where to from here?' that provides questions teachers could ask students, focusing on the big ideas involved at that level. This section provides new ideas to follow up and enables you and your class to enjoy yourselves thinking up new problems as extensions from the work of that level.

Additional resources are available at the series website, http://www.acer.edu.au/cam. Whenever these are referred to in the text, a globe icon appears in the margin. These include references and web links to related material, plus activity and summary sheets for students that provide a framework for their responses. These connect to the problems of that level and can be printed or copied for the students' use.

PART 1: NUMBER AND ALGEBRA

Part 1 presents three activities centred on the Number and Algebra strand.

Table 1.1: Number and Algebra activities

Problem	Big ideas
Number puzzles	• Addition of small numbers • Collecting and recording data • Using properties of numbers to continue patterns • Generalising from number properties and results of calculations
The Tower of Hanoi	• Collecting and recording data • Using properties of numbers to continue patterns • Generalising from number properties and results of calculations
Nim-like games	• Collecting and recording data • Using properties of numbers to continue patterns • Generalising from number properties and results of calculations

Some reminders before you use these tasks in your classroom:

1. The questions in the text are ones you can ask your students. You are likely to be able to produce similar, more immediately relevant ones for your particular students as you work on these activities with them.
2. We have suggested links to the Years in the Australian Curriculum: Mathematics for all of the Levels in each activity but, given that there will be a spread of ability in your classes, you should take these as a guide only. Take the opportunity to encourage every student to the edge of their comfort zone.
3. To take all students further, sometimes you can omit some of the later steps of a Level in favour of the early steps in the following Level.

CHAPTER 1: NUMBER PUZZLES

Initial problem

Is it possible to place the numbers 1, 2, 3, 4, 5 and 6 in the six circles such that:
» each number is used only once
» only one number is used in each circle
» the numbers on each of the three sides of the triangle all add up to the same total?

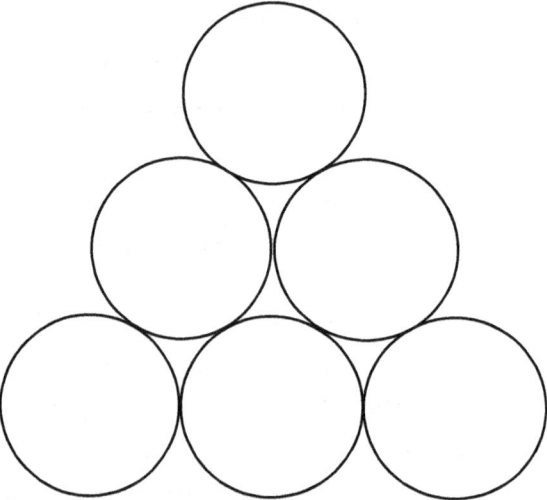

If it is possible, show it. If it isn't possible, then show that it can't be done.

Background information

This activity is a good example of how mathematics might develop. In the various steps we show how mathematicians slowly extend and generalise a problem. These are steps that most students can follow, if not necessarily undertake in full themselves.

It is also a good example of how different parts of mathematics can be combined to solve a problem. In the Level 1 problem we see how probability can be used, both to see what a conjecture might be and to check whether conjectures are on the right track.

Level 1 aims to help you scaffold your students through a solution to a problem including its proof. The initial problem is accessible to all students from Year 5 on, as it only requires knowledge of simple addition and guess-and-check. Only one creative idea (Proof 1 in Step 4) is needed, and it is worth challenging even the less able students to see how this works. Most students in Year 6 should be able to cope with this idea, though the algebraic proof at the end of this Level should be held until students reach Year 9.

The Olympic rings problem in Level 2 could be posed as a separate problem at any time. Many of the skills of Level 1 can be used here, but there are one or two developments that extend that work. For example, it is more efficient, though not necessary, if Proof 2 from Level 1 is used in the Olympic rings. What is more, there are multiple answers for some of the side sums. The initial problem of Level 2 is again accessible to Year 5 students and above. Many students in Years 7 and 8 may be able to solve the problem without using algebra. The whole solution given here should be left for Years 9 and 10 students.

Level 3 extends the idea of using the numbers 1 to 6, encountered in Level 1, to using *any* six numbers. Again, guess-and-check is a useful strategy. Looking for patterns in these sets is key, but not all of your students will be able to express these patterns algebraically. The early parts of Level 3 can be considered by most students in Years 7 to 8, but the general formulation of patterns, as well as the proofs, are for students with a good grasp of algebra.

Level 4 is a further extension of the Level 3 problem. Students in Years 7 to 8 with good algebraic skills can handle the Level 4 problem, though not all students need to tackle the proof there.

Table 1.2: Australian Curriculum content descriptions for the *Number puzzles* activity

Activity level	Problem	Content descriptions
1	Answers and proofs	*Year 7* Introduce the concept of variables as a way of representing numbers using letters (ACMNA175) Compare, order, add and subtract integers (ACMNA280) Assign probabilities to the outcomes of events and determine probabilities for events (ACMSP168) *Year 8* Simplify algebraic expressions involving the four operations (ACMNA192)
2	Olympic rings	*Year 7* ACMNA175 (see above) ACMNA280 (see above) *Year 8* ACMNA192 (see above)
3	Nice sets	*Year 7* ACMNA280 (see above) *Year 8* ACMNA192 (see above)
4	Finding all 'nice' sets	*Year 7* ACMNA280 (see above) *Year 8* ACMNA192 (see above)

Big ideas
» Describing and justifying patterns
» Using variables and algebraic expressions

Problem aims
» Interpret and solve word problems
» Think logically
» Extend the problem-solving strategies
» Generalise the outcomes from the solution

Key concepts
» Understand the relation between various parts of a problem in order to solve it
» Record and justify the answer using appropriate notation
» Use the basic arithmetic operations in a 'real' setting or context

Possible heuristics/strategies
» Guess and check (trial and error)
» Have I seen a problem like this before?
» Break a problem into manageable parts
» Test every possible combination

Concrete materials
» Numbered tiles
» Numbered pieces of paper

Special notes
» **If and only if:** This is a short way to represent two implications. In general, if A implies B and B implies A, we can say that A is true if and only if B is also true. A more detailed definition is provided on page 45 of this text.
» **Niceness:** When we use the word 'nice' in this book we mean something that arises that is initially unexpected and surprising but which has a nice feeling and seems to fit in exceptionally well.

Level 1: Answers and proofs

Problem

Is it possible to place the numbers 1, 2, 3, 4, 5 and 6 in the six circles such that:

- each number is used only once
- only one number is used in each circle
- the numbers on each of the three sides of the triangle all add up to the same total?

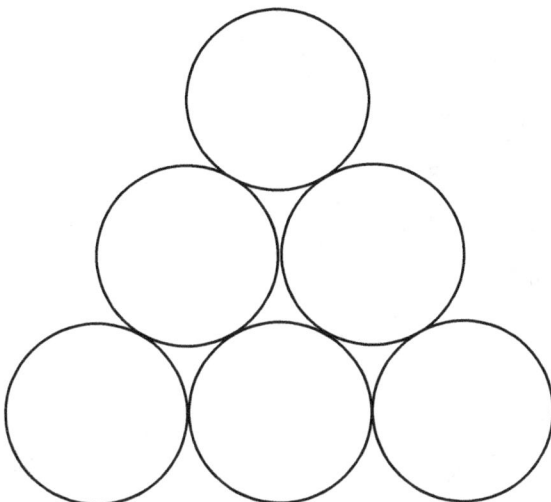

If it is possible, show it. If it isn't possible, then show that it can't be done.

Problem steps

Step 1

There are a few ways in which this problem might be tackled. The first is good old 'guess and check': just put numbers in and see what happens. This approach can be made systematic by first trying 1, 2 and 3 in the corners, then juggling around with 4, 5 and 6. Whether or not that works, a student might then try 1, 2 and 4 in the corners, and so on.

Another systematic way is to put 1, 2 and 3 along one edge of the triangle. If a student does this, they will soon realise that putting too many small or big numbers together causes an imbalance that won't let all the side sums be the same. But this may help them see a way to improve these trial and error processes.

Right at the start it is worth using a term such as *side sum* for the sum of three numbers on the edge of the triangle. Another approach might be to think about the possible side sums, then arrange the numbers to fit.

These attempts can be helped by using concrete materials such as numbered tiles or circular chips, or simply by putting the numbers 1 to 6 on separate small pieces of paper. The numbers can then be moved into position. This will be quicker than writing the numbers in circles all the time, but there is the disadvantage of not having a record of what has been done.

Whichever way students try to solve the problem, very few will try to show that there aren't any answers. It is always easier to get answers, if they exist, than it is to find proofs that they don't. So gradually the class will start to get answers. (These are shown in Figure 1.1 along with their side sums.)

Figure 1.1: Answers to the six-circle problem

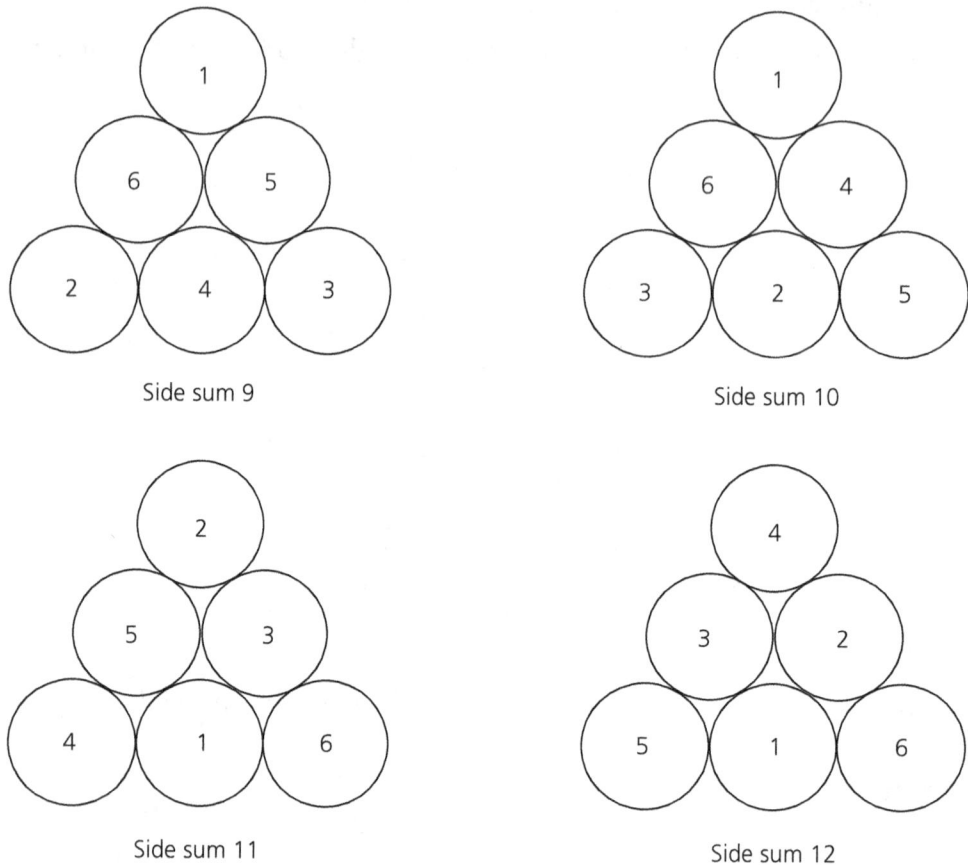

We can now work in one of two ways, depending on the class you have. If they all have similar mathematical abilities, let students put their answers on the board. This gives hints to the others about how the remaining answers might be achieved. It also shows that there are many ways to represent any of the solutions, because you can take any of the answers in Figure 1.1 and rotate or reflect them to produce six versions altogether (see Figure 1.2).

Figure 1.2: The various symmetries of the answer with side sum 9

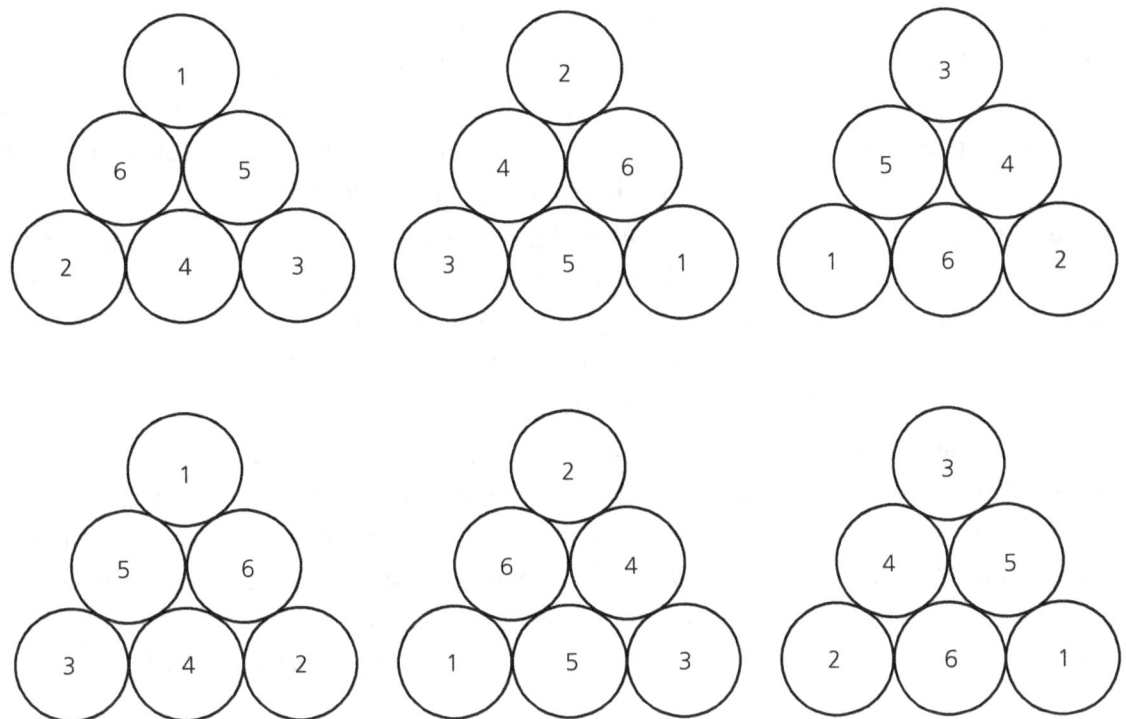

At some point, have the students vote on whether there are four or 24 solutions. Students may decide that these symmetric changes don't really give a new answer, so there are just the four possibilities in Figure 1.1. However, some classes may take the other stand and decide that there are 24 solutions. Either position is perfectly legitimate; it is up to you as to which your class works with. We assume there are only four answers from here on; however, we need to use 24 answers when determining probabilities in Step 2.

If there are significant differences in ability among your students, go round the class to monitor their work. When someone gets an answer, tell them to try to find more. If a student has found four answers, ask them to see if they can find any more; if they can't, ask them to prove that there are no more. Working this way should give every student the chance to find at least one answer, possibly with some scaffolding, before moving on.

Once a number of students have found four answers, have a vote as to whether there are only four answers or not. Those that believe there are only four answers have to prove this; those who think there are more have to find them. But it is likely that most students will settle on just four answers.

Conjecture 1: There are only four answers.

Step 2

We can either prove this conjecture or find a fifth answer. One way to test to see if we have a 'good' conjecture is to use probability.

Get each student to cut a sheet of paper into six small pieces and number them 1 to 6. They then put the pieces into a box or container (or at least put them face down and mix them around) and then randomly draw them out. As each piece is drawn, the student writes that number in some pre-specified order into the six circles. They then check whether they have an answer (all side sums the same) or not. If they get an answer that is not one of the four established possibilities, this means there is a fifth answer and the conjecture has to be changed. If it isn't a new answer, they keep going. Have them keep doing this until the class accumulates well over 100 trials overall. By looking at the number of successful trials, students can get a rough idea of the probability of finding a reasonable answer.

Now compare this to the predicted probability of getting an answer at random if there are four answers. There are six ways to choose the first number to put in any position. That leaves five ways to put the remaining five numbers in the circles left, then four ways to place the remaining four numbers, and so on. This means that altogether there are $6 \times 5 \times 4 \times 3 \times 2 \times 1 = 720$ possible outcomes of the trials.

On the other hand, if there are only four answers, each one would come up in six ways because of symmetry. So there are $4 \times 6 = 24$ possible successful trials, which leads to a probability of $\frac{24}{720} = \frac{1}{30}$. If the experimental probability is around $\frac{1}{30}$, then it is likely that there are only four answers and they should concentrate on a proof. On the other hand, if the experimental probability hovers around the $\frac{30}{720} = \frac{1}{24}$ mark, then your students should be looking for a fifth answer. (Of course if it is $\frac{1}{20}$ there might be six answers, and so on.)

Step 3

Before going on to a proof, it is worth looking at the answers more carefully, first in small groups and then as a whole class, and discussing anything interesting that students might find. The following ideas are not listed in any particular order.

- The only side sums seem to be 9, 10, 11 and 12. Are these numbers an accident? Why aren't there any smaller sums or any larger ones? We will say more about this in Step 4.
- The numbers in the corners are either consecutive numbers or differ by 1 or 2. Can any other situation occur? Again, see Step 4.
- The middle numbers of one answer become the corner numbers of another, and vice versa. You can check this by looking at the links between the ones with side sums 9 and 12, and 10 and 11.
- Think of the numbers of an answer as being on a rubber belt. If you pull the belt around so that the numbers move on one circle, you get another answer.
- Most surprising of all, if you change any number n to $7 - n$ (so 1 goes to 6, 2 to 5, 3 to 4, 4 to 3, 5 to 2 and 6 to 1), the resulting numbers form an answer as well. Check it out.

There may be other things of interest to students and these should be discussed as well. All of this should give your students a better feel of the problem and will be valuable in the remaining levels of this activity.

Step 4

Now is the time to look for a proof. Ask each student group to look seriously at Conjecture 1 and see how it might be proved. If they get a proof, they need to convince the class that what they claim is true.

It turns out that there are two common proofs. The first comes from the need to balance the numbers so that too many big numbers do not go together. The second is essentially an algebraic idea (though it might not necessarily be stated that way initially).

If a proof doesn't seem to come easily for the class, then scaffold whichever you think they can understand most easily. On the other hand, if you think that a proof is beyond them, move on to Level 2.

Step 4.1

Proof 1: This concentrates first on the side sums and relies on a simple idea: '*6 has to go somewhere*'. So what is the smallest side sum that uses the 6? The smallest numbers to add to the 6 are 1 and 2. So the smallest possible side sum is 6 + 1 + 2 = 9.

Surprisingly, it may take a while before anyone comes up with the inverse: '*1 has to go somewhere*'. So what is the biggest side sum possible? 1 + 5 + 6 = 12.

We now know that we can't have side sums below 9 and above 12, and we also know that we can have at least one each of 9, 10, 11 and 12. But it is not obvious that there is only one answer for each side sum. We show an argument for 9 below; the cases for 10, 11 and 12 follow in the same way. (If your students need you to scaffold them for the 9 case, let them work out one of the other cases for themselves.)

$$9 = 6 + 2 + 1 = 5 + 3 + 1 = 4 + 3 + 2$$

These are the only ways to get 9 and these can be found by being systematic. (First say that if we use 6, we need to make up 3 and this can only be done by adding in 2 and 1. Now we are not allowed to use 6 any more. If we use 5, we need to make up 4 and this can only be done with 3 and 1. Now we can't use 6 or 5. With 4 we need to make up 5 and this requires 3 and 2.)

There is a good chance that someone will notice that 1, 2 and 3 occur twice in the three sums for 9. That has to mean they must be used on the corners of the triangle, so that they are counted in two side sums. Once you put these on the corners, the middle numbers fall into place to give the answer shown in Figure 1.1.

The really creative idea here is to think of the '6 has to go *somewhere*' argument.

Step 4.2

Proof 2: This is an algebraic proof. You might encourage students to try algebra, especially if you want to show that algebra is useful for things other than solving linear equations. Here they use it first to find the possible side sums and then to complete the answers.

Figure 1.3: The start of an algebraic approach

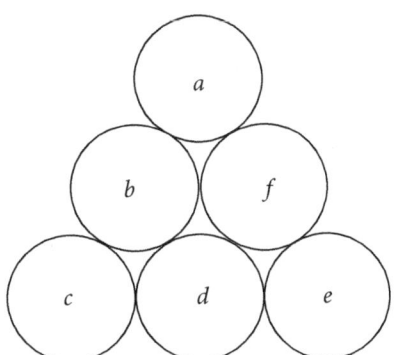

Put the variables in as shown in Figure 1.3. The next step is to add the three sides together. So:

$$(a + b + c) + (c + d + e) + (e + f + a) = \text{three times the side sum}$$

But $a + b + c + d + e + f = 1 + 2 + 3 + 4 + 5 + 6$ (where the digits aren't necessarily in that order to start with). So $a + b + c + d + e + f = 21$. This means that:

$$(a + c + e) + a + b + c + d + e + f = (a + c + e) + 21 = \text{three times the side sum}$$

The smallest value the corner numbers can have is $1 + 2 + 3 = 6$, so three times the smallest side sum is 27. We can't get a side sum smaller than 9. (That is refreshingly consistent with Proof 1!)

But the biggest numbers we can have in the corners are 4, 5 and 6, so three times the smallest side sum is 36. Our biggest side sum is 12.

Now we know what the corner numbers should be and what side sums they correspond to, we can just finish off all four answers.

The big creative ideas here are not just to think of using algebra, but to link it to the side sum and to realise that the sum of all the letters is 21.

Where to from here?

- On a scale of 1 to 10, with 10 being the best problem they have seen, what do your students think of that problem?
- On a scale of 1 to 10, with 10 being the 'nicest' piece of mathematics they have seen, what do your students think of the maths used here?
- What do they think was the 'nicest' idea in this problem?
- Now they have seen this problem, can they make up other problems that might use the same ideas?

Level 2: Olympic rings

Problem

Is it possible to fit the numbers 1 to 9 in the spaces in the Olympic rings, so that the 'straight line' sums $a + b + c$, $c + d + e$, $e + f + g$ and $g + h + i$ are all the same?

Figure 1.4: The Olympic rings problem

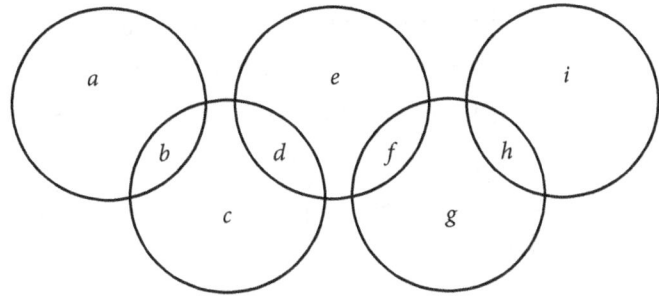

Problem steps

Step 1

While this problem is similar to the six-circles problem, there are sufficient points of difference to make it worth solving.

Start by discussing this problem with your class. Get them to realise that they have seen a problem like this before. Encourage them to make a plan of attack on this problem, then let them work in groups to work through this plan. The main steps in the plan might resemble this list.

1. Find what the 'side sums' could be.
2. Find answers for each of the side sums.
3. Find all answers for each side sum.
4. Prove that all answers have been found.
5. Look at any overall structures and patterns that exist in their results.
6. Extend the problem to finding all possible sets of nine numbers that give answers.
   Now let the class experiment.

Step 2

First, students should consider the possible side sums. Two methods are shown in Level 1, Step 4. Proof 1 shows that side sums of from 12 to 19 are possible, while Proof 2 gives the possibilities of only 13 to 17. It is probably time to wean your students off Proof 1, as from this point it is less efficient than Proof 2. The problem is that the special cases, like the 12 and 19 side sums, will get more painful to rule out as we progress through the levels. Encourage students to understand and use Proof 2.

(However, it is better for students to make progress on a problem than to be forced into the algebra of Proof 2. You will have to make that judgement for individual students.)

So Proof 2 shows that $(c + e + g) + (1 + 2 + 3 + 4 + 5 + 6 + 7 + 8 + 9) = 4(\text{side sum})$; or

$$4(\text{side sum}) = 45 + (c + e + g)$$

Now the smallest side sum might use c, e, g as 1, 2, 3, but $45 + 6 = 51$ isn't a multiple of 4. We have to move to 52 to get a multiple of 4, and if $4(\text{side sum}) = 52$ then the side sum is 13.

On the other hand, the biggest side sum would be found if c, e, g was 7, 8, 9, but we get the same divisibility problem that we found with 1, 2, 3. The largest number less than 69 that is divisible by 4 is 68, and if $4(\text{side sum}) = 68$, side sum = 17.

As a result of this, any side sums that exist must be in the range 13 to 17.

Let the groups discuss their results with the class. Ask them which of Proofs 1 or 2 do they think is the most efficient here. Make sure they see that, although Proof 1 is in some sense the easiest to use because it doesn't require algebra, Proof 2 saves a lot of work going through extra cases.

Step 3

Let the class play around with a side sum of 15 to see how many answers they can get. They will soon conjecture that there aren't any. They should see that to get a side sum of 15, the corner numbers have to add to 15 as well. But if $c + e + g = 15$, then d will have to equal g. So there are no answers with a side sum of 15.

Step 4

From here, lead the class to looking at the number of answers that can be found for the side sum of 13. After some guessing and checking they should conjecture that the only answer is the one found in Figure 1.5.

Figure 1.5: An answer for the side sum of 13

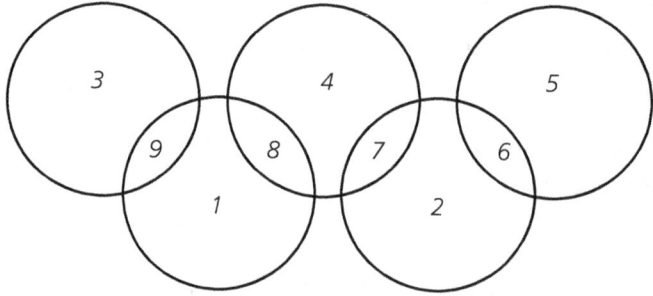

To show this is the only one, students first need to note that the corner spaces (c, e and g) have to contain 1, 2 and 4. But 1 and 2 can't be in the same sum, because they would need to have a 10 to make up the 13. So 1, 2 and 4 have to lie in the positions shown in Figure 1.5. The rest is straightforward because 1 and 4 need an 8; 2 and 4 need a 7; the 9 has to go with the 1; and the remaining numbers are then forced.

Note that, as in the six-circles problem, reflective symmetry through the centre circle produces another possible solution from any given solution. But the Olympic rings give a further possibility. In the case of the answer in Figure 1.5, for example, we can interchange the 3 and the 9 and the 5 and the 6 to give something that is apparently different (see Figure 1.6). However, we will assume that both of these 'twists' give the same result as the answer in Figure 1.5, although it is valid to say that they produce eight answers in total.

Figure 1.6: An alternative answer for the side sum of 13

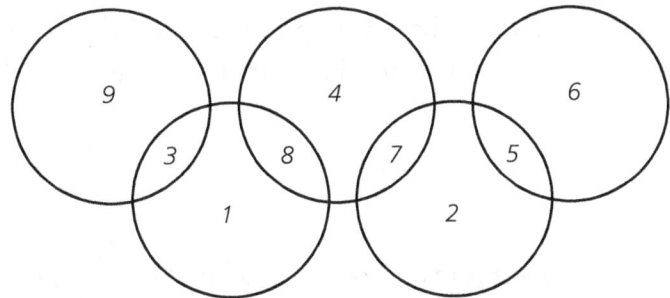

While we are here, we can adapt a method from Level 1. If we interchange 1 and 9, 2 and 8, 3 and 7, 4 and 6, and 5 and 5 (that is, go from n to $10 - n$) we get an answer with a side sum of 17. Because the same interchange will get us back from 17 to 13, there must be only one 17. Otherwise we would have found more 13s.

(A proof that this interchange works: observe that changing n for $10 - n$ changes a side sum from s to $30 - s$. This is demonstrated by $30 - 13 = 17$ and $30 - 17 = 13$.)

Step 5

Students should now work on the side sum of 14. This has more possibilities.

Groups may decide that there are four possible answers. Proving this is a bit messy. It involves first finding all the ways to get 14 (the sum of the three corner numbers using Proof 2). There seem to be eight of these, though half of these fail in much the same way as $9 + 4 + 1$, which we will now explain. 1 and 4 can't be on the same side, as they need a 12 to get to 17. So let $c = 1$, $e = 9$ and $g = 4$. Then d has to be 7 and f has to be 3. To complete the final two sides we have to repeat a number.

As a result of the last part of the previous level, there are four answers with a side sum of 16.

Where to from here?

- Ask your students to list all of the techniques that can be applied to the two problems of Levels 1 and 2.
- Ask the class to write a series of steps (an algorithm) that would enable them to find all the answers to a problem of this type. There are several ways of writing such an algorithm by building in logical checks, such as the limits to the side sums. One strategy is to test every possible combination of the nine digits. This involves testing 9! (= 388 000) combinations of positions, which seems (and is) very inefficient, but the speed of modern computers can produce all of the answers in a very short time.
- What problems can the class invent that are similar to the ones in Levels 1 and 2?

Level 3: Nice sets

Problem

You can call a set of six whole numbers *nice* if you can put the numbers into the six circles of the six-circles problem (Level 1) so that the numbers on each side add up to the same sum.

Find as many *nice* sets as you can. Do they all produce four arrangements, as {1, 2, 3, 4, 5, 6} did?

(Note: this problem evolves from the Level 1 problem by asking the simple mathematical question 'what if?': What if we can use numbers other than just 1 to 6?)

Problem steps

Step 1

Usually, the nice sets that students find are consecutive numbers, starting with {2, 3, 4, 5, 6, 7} and moving upwards. To show that this set works, it is only necessary to add 1 to each of the numbers in the solution.

The common property here is consecutive numbers. So $\{n, n + 1, n + 2, n + 3, n + 4, n + 5\}$ covers all nice sets. As before, we can make up four possible arrangements by returning to Figure 1.1 (p. 14) and replacing 1 by n, 2 by $n + 1$, 3 by $n + 2$, 4 by $n + 3$, 5 by $n + 4$ and 6 by $n + 5$. This tells us that are an infinite number of nice sets.

Step 2

Students often see multiples next. So {2, 4, 6, 8, 10, 12} can be used in the same way we used consecutive numbers. It then follows that we can get four successful arrangements for $\{a, 2a, 3a, 4a, 5a, 6a\}$ where we replace the numbers {1, 2, 3, 4, 5, 6} by the same multiple of each.

Step 3

Some students will then combine these to get nice sets such as {5, 7, 9, 11, 13, 15}. In its general form this is $\{a + b, 2a + b, 3a + b, 4a + b, 5a + b, 6a + b\}$. For the example above, $a = 2$ and $b = 3$.

We again get four arrangements. This is not surprising, because the sets in Step 3 contain all the sets in Steps 1 and 2. For Step 1 answers, put $a = 0$ and $b = 1$; for Step 2, put $a = 1$ and $b = 0$.

About this time, an advanced student may branch out into negative numbers or even fractions or decimals. Clearly {−15.35, −10.35, −5.35, −0.35, 4.65, 9.65} will work. Encourage them to use any real numbers they like. If nothing else, they will get a chance here to practise arithmetic with fractions and decimals.

Step 4

Ask the students if they have any conjectures about nice sets. We suggest the following two.

> **Conjecture 2:** The only nice sets are of the form $\{a + b, 2a + b, 3a + b, 4a + b, 5a + b, 6a + b\}$.

> **Conjecture 3:** Given any nice set, there are precisely four answers it can produce.

What progress can they make with these two conjectures?

Step 5

If your students have not produced any proofs or counter-examples to the question at the end of Step 4, ask them to think about $\{1, 2, 3, 4, 5, n\}$. For what n is this set nice?

We know that $n = 6$ gives a nice set. And it certainly looks as if it won't be a nice set if $n = 1000$; the 1000 can't be balanced up by the other numbers. But are there any values of n between 6 and 1000 that will give a nice set?

The general student response is that there can't be, but get them to persist. What about $n = 7, 8$ or 9?

This can be approached using either of the proofs of Level 1, Step 4. We'll use the Proof 1 of Step 4.1.

Since 7 has to be there, the lowest sum is $7 + 2 + 1 = 10$ and the highest is $7 + 5 + 1 = 13$. This gives possible side sums of 10, 11, 12 and 13. If we use the systematic approach to get these sums, we find:

- $10 = 7 + 2 + 1 = 5 + 4 + 1 = 5 + 3 + 2$
- $11 = 7 + 3 + 1 = 5 + 4 + 2$ and no others
- $12 = 7 + 4 + 1 = 7 + 3 + 2 = 5 + 4 + 3$
- $13 = 7 + 5 + 1 = 7 + 4 + 2$ and no others.

Two of these give us arrangements; see Figure 1.7.

Figure 1.7: The surprisingly nice set $\{1, 2, 3, 4, 5, 7\}$

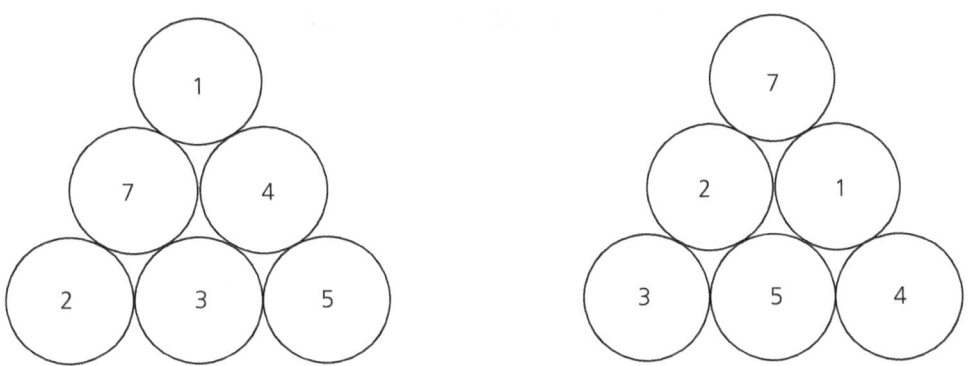

This, of course, has made a big hole in the last two conjectures!

But what about {1, 2, 3, 4, 5, 8}? Taking the approach on page 23, we see that the possible side sums are 11, 12, 13 and 14. Working systematically we get:

- $11 = 1 + 2 + 8 = 5 + 2 + 4$ and no others
- $12 = 1 + 3 + 8 = 5 + 3 + 4$ and no others
- $13 = 1 + 4 + 8 = 2 + 3 + 8$ and no others
- $14 = 1 + 5 + 8 = 2 + 4 + 8$ and no others.

None of these give three sums and so we don't have another nice set. The same can be shown for {1, 2, 3, 4, 5, 9} and {1, 2, 3, 4, 5, n} for $n > 9$.

Step 6

Does {1, 2, 3, 4, 5, 7} lead to an infinite number of nice sets? What can your students come up with?

The same infinite variations that were found in Level 1 will work here. So any set of five consecutive numbers, plus the number that is two bigger than the largest of these consecutive numbers, will be nice. Any multiple of {1, 2, 3, 4, 5, 7} such as {3, 6, 9, 12, 15, 21} will also work.

Step 7

Some students may also try something like {1, 2, 3, 6, 7, 8}. This only has two arrangements, one with side sum 11 and one with side sum 16.

However, this type of nice set—{$a, a + c, a + 2c, b, b + c, b + 2c$}—may not appear until you ask some of the questions in Level 4.

Where to from here?

- What happens if we replace whole numbers by fractions or decimals?
- What conjectures did your students come up with now? Which of these could they prove?
- Can your students find nice sets for the Olympic rings of Level 2?
- What did they think was the nicest thing in this level? Did anything make them uncomfortable?
- What extensions of the problems so far can they invent?

Level 4: Finding all nice sets

Problem

Suppose that someone gives you some (less than six) numbers. Can you *always* find a nice set that contains the given numbers? Can you find an infinite amount of numbers to complete the nice set?

Problem steps

Step 1

Discuss the problem with the class as a whole. Give them one number, say, 23. Can they now find five other numbers to make up a nice set?

This shouldn't be too hard. We know from Level 2, Step 1 that consecutive numbers make a nice set, so {23, 24, 25, 26, 27, 38} will do. But from Level 3, Step 3 (page 22), we know that $\{(23 - r) + r, 2 \times (23 - r) + r, 3 \times (23 - r) + r, 4 \times (23 - r) + r, 5 \times (23 - r) + r, 6 \times (23 - r) + r\}$ is nice for any number r. So that gives us an infinite set of nice sets using 23.

Students may come up with many other kinds of infinite nice sets using the tricks (logic) of Level 1. Many of these may be much simpler than the example we have chosen. Spend a little while listening to students' thoughts on their processes, as it might help them in the rest of this problem.

Step 2

Graduate from giving them one number to two, such as 6 and 7, and ask student groups to find an infinite number of nice sets that contain 6 and 7.

Since 6 and 7 differ by 1, it is easy to use the consecutive-number nice set to get one such, but there are many others. Level 3, Step 7 can give you an infinite number of nice sets. Here is one example: $\{6, 6 + a, 6 + 2a, 7, 7 + a, 7 + 2a\}$, where a is chosen so that all numbers in the set are distinct.

Step 3

Ask students what *they* think you are going to ask them next. Clearly you want to know if they can make a nice set from any three different numbers.

Let them do some specific examples first. These problems can be done by a suitable use of Level 3, Step 7. Take the two smallest numbers to be the smallest numbers in the set, and let their difference be a. Let the biggest number be the c of the nice set.

Step 4

With four given numbers, things get more serious. Give them any four numbers that you can think of. We suggest 1, 2, 8 and 9 to start with, followed by 1, 2, 8 and 10.

For the first of these examples, $\{1, 2, 8, 9, x, x + 1\}$ will do for any x or $x + 1 \neq 1, 2, 8$ or 9. For the second example, $\{1, 2, 8, 9, 10, 11\}$, $\{1, 2, 8, 9, 10, 17\}$, $\{1, 2, 8, 10, 11, 17\}$ and one or two others should work. But is there an *infinite* number of nice sets containing 1, 2, 8 and 10?

Step 5

While some sets of five numbers will lead to a nice set, will *any* five numbers do so? How about 1, 2, 3, 4, 1000?

If $\{1, 2, 3, 4, 1000, x\}$ is a nice set, then using the idea of Proof 2 we know that:

$$(a + c + e) + 1010 + x = 3(\text{side sum})$$

But the side sum has to be at least 1000, so the right side has to be at least 3000. The left side can never reach these heights. This makes it clear that not every set of five numbers can lead to a nice set.

Step 6

So how do we know if a set of six numbers is nice? Do your students have any conjectures? They may have noticed that the differences between a corner number and the number on the side opposite that corner number are all the same.

We now prove the following theorem.

> **Theorem:** The set $\{a, b, c, d, e, f\}$ is nice *if and only if* the numbers can be arranged in three pairs, each of which has a common difference.
>
> **Proof:** Suppose that $\{a, b, c, d, e, f\}$ is a nice set. Then from Figure 1.3 on page 17 we can see that:

$$a + b + c = c + d + e$$

so

$$a - d = e - b$$

and

$$c + d + e = e + f + a$$

so

$$c - f = a - d$$

This means that $a - d = c - f = e - b$. Call these differences Δ.

So $\{a, b, c, d, e, f\} = \{b, d, f, b + \Delta, d + \Delta, f + \Delta\}$. This gives three pairs: $b, b + \Delta$; $d, d + \Delta$; $f, f + \Delta$, with the difference between the numbers in the pairs all being the same.

Now suppose that we have pairs $p, p + \varepsilon$; $q, q + \varepsilon$; and $r, r + \varepsilon$. They form a required arrangement as shown in Figure 1.8, where the arrows indicate the side sums of each side.

Figure 1.8: The nice set {$p, q, r, p + \varepsilon, q + \varepsilon, r + \varepsilon$}

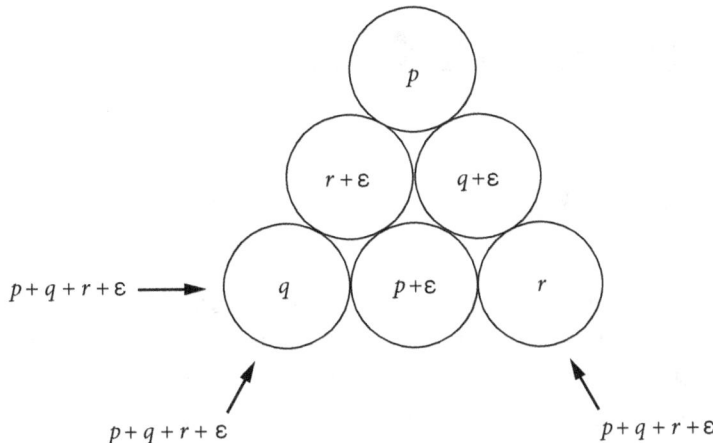

From the theorem, we have a simple test for deciding whether or not a set is nice.

Where to from here?

- One direction we have not yet taken is to completely determine the number of answers for a given nice set. We will leave these conjectures for your students to think about.

    Conjecture 4: A nice set has to have an even number of answers.

    Conjecture 5: A nice set has four answers *if and only if* it is of the form {$na + b$, $(n + 1)a + b$, $(n + 2)a + b$, $(n + 3)a + b$, $(n + 4)a + b$, $(n + 5)a + b$}.

    Conjecture 6: If a nice set is not of the form in Conjecture 5, then it has only two answers.

- Can they determine whether or not the conjectures are true?
- What did your students think was the most interesting part of this level?
- Can the class think of some other problem? How far can they take it? There is no need to prove any of this; just let them try to extend the problem as far as they can.
- An extension problem, 'Polygonal puzzles', can be downloaded from the series website.

CHAPTER 2:
THE TOWER OF HANOI

Initial problem

A shrine in a remote place has three vertical ivory pegs, along with 64 golden discs that fit onto the pegs. The discs all vary in radius, with the largest one on the bottom of the pile, the next largest on top of that, and so on up to the smallest on the top. (In Figure 1.9 below we show a picture of a Tower of Hanoi model with just four discs on it.)

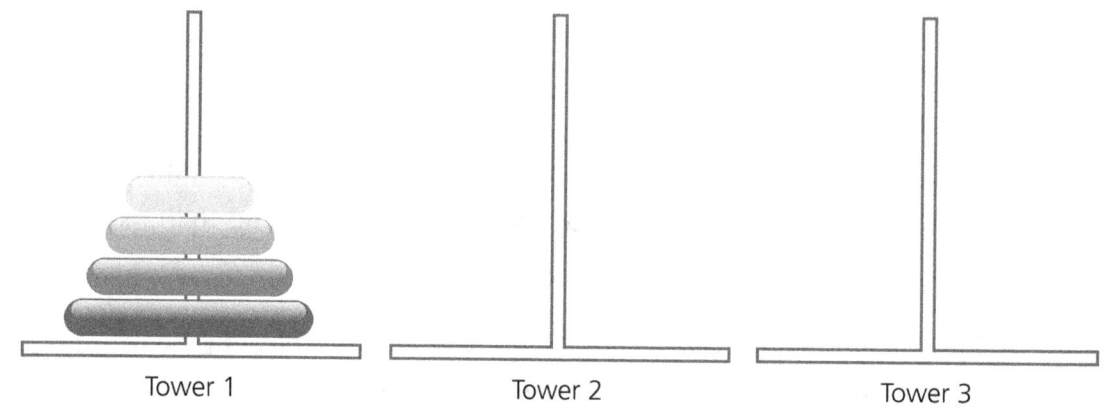

Figure 1.9: The Tower of Hanoi game with four discs

At the start of time, the discs were all on one peg: the left-most peg. Every day, the monks of the shrine move one disc onto another peg. But there is a rule and a goal. The rule is that any newly moved disc must be placed either on an empty peg or on top of a larger disc. The goal is to move all of the discs onto the right-most peg so that the largest one is on the bottom, the next largest on top of that, and so on up to the smallest on the top.

Legend has it that the end of the world will come when the goal is achieved. How long will the world last?

Background information

This problem seems to have been invented in 1883 by the French mathematician Édouard Lucas. (We say 'seems' because some sources suggest that Lucas discovered and publicised a much earlier version of the problem.)

The main aim of this activity is for students to see and experience the famous *Tower of Hanoi* problem. However, it is also valuable for finding patterns through guessing (conjecturing) and justifying these.

Each group of students must have access to a Tower of Hanoi model. These can be purchased online and from educational supply companies; there are also applets and simulations online that students can use (see links on the series website). As an alternative, it is not difficult for students to make their own models using cardboard.

We suggest that they cut out six squares (discs) of varying sizes, then a rectangular base on which three layers of squares will fit. This can be done using two rectangles, as shown in Figure 1.10. Students cut three square holes in the rectangle to represent the three pegs, then glue the two rectangles together. (More ambitious students may even want to insert real pegs into their model.)

Get each student in a group of three or four to take turns moving one of the discs. Don't let one student take over the model.

Clearly, 64 discs is far too many to start with, but a Tower of Hanoi model with just a few discs is a good starting point to get initial success. You can then use it as a springboard for the more complex situations of 4, 5, 6 and more discs.

Once again, this activity moves through its levels by gradually working at and solving specific special cases. This path leads to a solution of the initial problem. Much of the first activity is accessible to all students through a guess-and-check approach. This process should give many students the chance to find some reasonable conjectures and most of them will be able to appreciate the proofs, if not find them for themselves.

In Level 1 we look at the special cases of one to four discs. As an introduction, Level 1 is open to all students as they play the game.

This is extended to eight discs in Level 2. Level 2 is also accessible to all students, although it requires the students to have some facility with powers of 2. The little algebra used here could be avoided by concentrating on special cases.

Working through the first two levels should give students enough understanding to make the conjectures of Level 3. Some Year 7 students will be able to do Level 3, but more success will be gained by older students who are more familiar with algebra.

While the first three levels concentrate on the number of moves, Level 4 aims to promote an understanding of what moves should be made at each stage of the play. This enables students to provide an algorithm that could theoretically allow a computer to do the hard work for them. Level 4 requires more insight than algebraic ability, so with help it is open to all students; however, it does require more mathematical knowledge than the previous Levels and so is best used with students from Years 9 and 10.

Table 1.3: Australian Curriculum content descriptions for the *Tower of Hanoi* activity

Activity level	Problem	Content descriptions
1	Four discs	
2	Patterns	*Year 7* Investigate index notation and represent whole numbers as products of powers of prime numbers (ACMNA149)
3	Proofs	*Year 7* ACMNA149 (see above) Introduce the concept of variables as a way of representing numbers using letters (ACMNA175) *Year 8* Simplify algebraic expressions involving the four operations (ACMNA192) *Year 9* Apply index laws to numerical expressions with integer indices (ACMNA209)
4	How to move the discs	*Year 9* ACMNA209 (see above)

Big ideas

» Understanding problems
» Finding patterns (by inductive reasoning)
» Justification (proof)
» Communication

Problem aims

» To find and justify patterns found in a game

Key concepts

» Powers of 2

Resources

» Tower of Hanoi model or applet
» A calculator that will give you answers to more than 20 decimal places

Possible heuristics/strategies

» Simpler (smaller) cases
» Make a table
» Be systematic
» Break a problem into manageable parts

Level 1: Four discs

Problem

A shrine in a remote place has three vertical ivory pegs, along with 64 golden discs that fit onto the pegs. The discs all vary in radius, with the largest one on the bottom of the pile, the next largest on top of that, and so on up to the smallest on the top. (Below we show a picture of a Tower of Hanoi model with just four discs on it.)

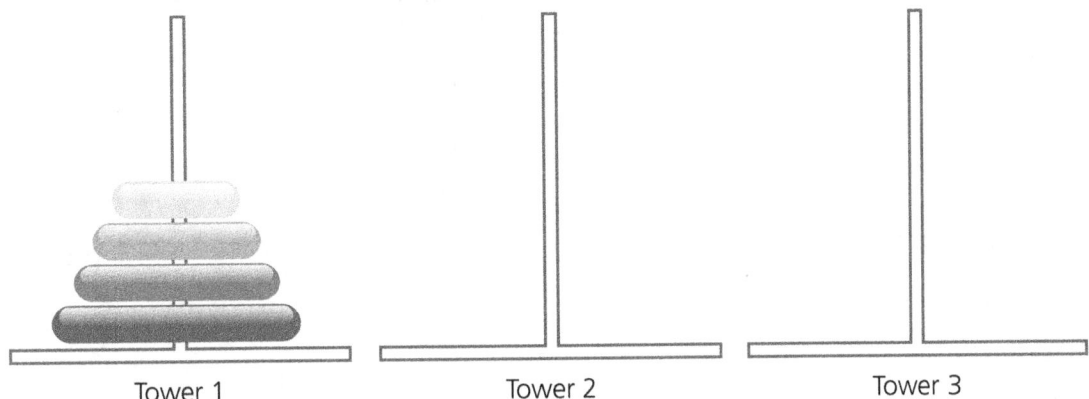

At the start of time, the discs were all on one peg: the left-most peg. Every day, the monks of the shrine move one disc onto another peg. But there is a rule and a goal. The rule is that any newly moved disc must be placed on an empty peg or on top of a larger disc. The goal is to move all of the discs onto the right-most peg so that the largest one is on the bottom, the next largest on top of that, and so on up to the smallest on the top.

Legend has it that the end of the world will come when the goal achieved. How many moves are needed? How long will the world last?

Problem steps

Step 1

The original problem is far too hard for most students to tackle from scratch. However, it is worth starting by discussing the steps of the puzzle with the class and how long it might take the monks to move the discs. Record these times and come back to them later. (The answer is about 5×10^{16} years!)

A useful problem-solving strategy is to use a *simpler case*. At this level we try two, three and four discs (the number of moves for one disc is obvious).

Step 2

Ask the students if it is possible for the monks to move *two* discs from one peg to another.

There are two purposes for doing such a simple step. First, it gives the students a chance to understand the problem and the moves required. Second, it gives the students the number of moves for two discs, which will be valuable later in Step 3.

Figure 1.10 shows that three moves are needed.

Figure 1.10: The moves needed where the original peg has two discs

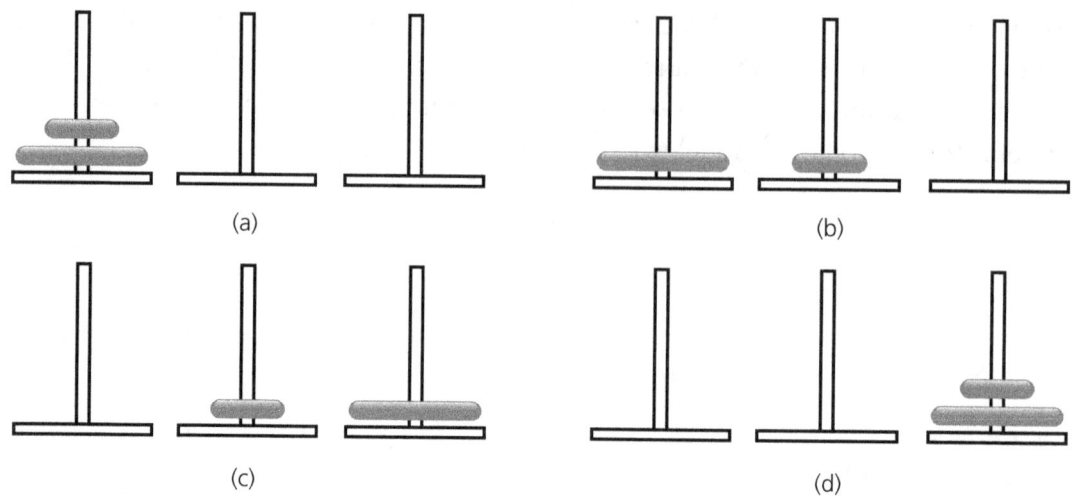

Step 3

Ask the class if it is possible for the monks to move three discs.

Let the class experiment to see what they can find, using a model or applet. You may want to let them do this more than once to make sure they get the right answer. If they are struggling, only provide a limited degree of help, such as reiterating the rules, asking what happens if they do this or this, asking them what has gone wrong, or suggesting that they try going back to one or two discs to get the rules correct. Also, let each student in a group take turns to move a disc. This prevents students being excluded from playing with the equipment and thus being obstructed in developing their intuition on this problem.

Don't record the moves as the students are working on this. However, when you are happy that they have the basic ideas involved, let them draw a poster of each of the moves from start to finish. Figure 1.11 uses *seven* moves to reach the correct answer. Don't intervene at this point if some posters show more moves than are really required.

Figure 1.11: The moves needed where the original peg has three discs

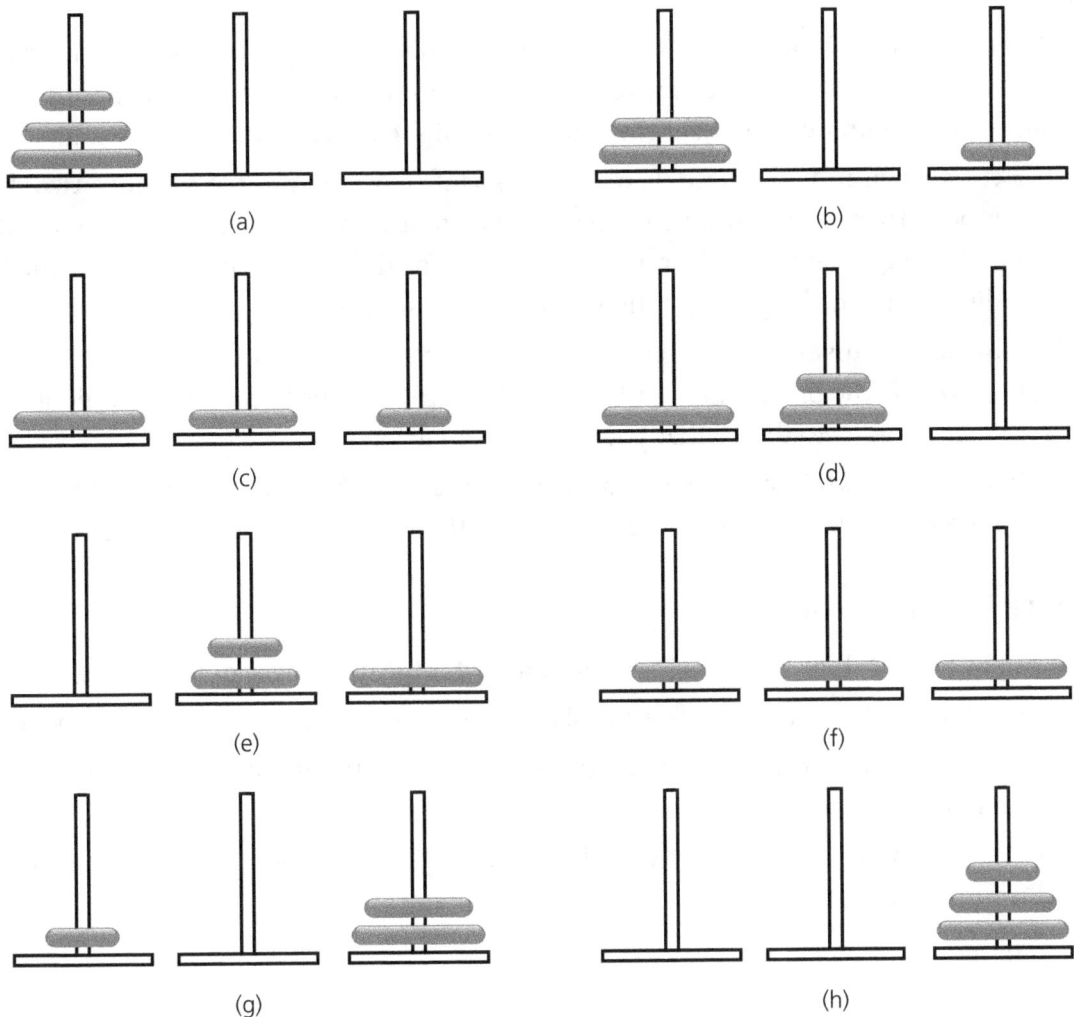

At this point one of two things might happen. Students might see that they can successfully move three discs, or they may see that to move three discs they first moved the top two discs (see moves (b), (c) and (d) in Figure 1.11). In either case, don't react strongly but just continue to the next step.

Step 4

Now ask the class what is the *smallest* number of days that the monks need to move three discs.

Get them to discuss this first in their groups and then as a whole class. In this discussion they might see that some people have taken too many moves in their posters. If no one has got as few as seven moves, scaffold them to get that number. You can do this by pointing out the posters that have the fewest moves so far and asking them to consider what alternatives there might be before each move. (This is the beginning of getting them to *justify* their answer.) Some students may now want to make some changes in their poster. When this has been done, let the class choose a couple of the posters for display in the classroom.

At this point the students may not have totally justified that the fewest number of moves is seven. This can be returned to later.

Ask the class what they think they should do next with this problem.

Step 5

Ask the students if it is possible to move four discs from one peg to another. What do they think is the *smallest* number of moves needed for four discs? If possible, these guesses should be backed up by a sound argument. What might such an argument look like?

In Step 3, when moving three discs, it looks as if they first needed to move two discs to a new peg. Then they moved the biggest disc, followed by moving the two smallest discs onto the biggest disc. So the number of moves for three discs could be: 3 for the two smaller discs, 1 for the biggest disc then 3 for the two smaller discs.

In moving four discs it seems reasonable that they might first move the three smaller discs, followed by the biggest disc, followed by the three smaller discs. This would give us 7 + 1 + 7, which adds to the 15 that your students may have already found.

Is there a generalisation here? Is it always the case that this method works? So for 64 discs, do the monks have to move the top 63 first, then the largest disc, then the others?

Where to from here?

- It might be useful for the class to produce a poster for the moves of four discs.
- Make a movie of the moves for five and six discs. Does this help to see what is happening?
- What do they think the number of moves for five discs might be? How many moves might it take to move 64 discs?
- Ask the class to think of further questions that might be considered here. Let them follow up some of these.

Level 2: Patterns

Problem

Are there any patterns in the numbers of moves for the *Tower of Hanoi* problem?

Problem steps

Step 1

Students generally have little trouble coming up with the three conjectures below and their patterns. However, they may not see them in the order given.

First of all, we have the conjecture that was taking shape at the end of Level 1:

Conjecture 1: The number of moves for d discs is twice the number of moves for $(d-1)$ discs, plus 1.

If this is true, it means that we can extend our known results for 1, 2, 3, and 4 discs to 5 and 6 discs and beyond. We have highlighted the moves for 5 and 6 discs in Table 1.4.

Table 1.4: The probable moves required for small numbers of discs

No. of discs	1	2	3	4	5	6
No. of moves	1	3	7	15	31	63

It might be worth seeing if Table 1.4 gives the correct value for five discs by trying to get fewer than 31 moves there.

Now ask the class to calculate the number of moves for 64 discs based on Conjecture 1. They will quickly find that this is tedious and open to errors of calculation. It would be better if we could find a formula for any number of discs, and then just put 64 in the formula. Is there another approach for finding this formula?

Step 2

We could look for other patterns in the numbers that might help us. Students often like to see what the differences are between consecutive terms, which should lead them to Table 1.5.

Table 1.5: The probable differences between moves

No. of discs	1	2	3	4	5	6
No. of moves	1	3	7	15	31	63
Differences		2	4	8	16	32

It looks as if the numbers are going up by a power of 2. For example, the difference between two discs and one disc is 2; between three discs and two is 4; and between four discs and three is 8.

But what is the power of 2 in each case? For two discs it goes up by $2 = 2^1$; for three discs it goes up by $4 = 2^2$; for four discs it goes up by $8 = 2^3$; and so on. It looks as if the power of 2 is one less than the number of discs. That gives us the next conjecture.

Conjecture 2: The number of moves for d discs is 2^{d-1} more than the number of moves for $(d-1)$ discs.

Ask your class if they want to find the number of moves for 63 discs based on this conjecture. This would be just as hard as trying to find that answer using Conjecture 1.

Step 3

Can they take a frontal attack here? Can they see any pattern in the numbers of moves in Tables 1.4 and 1.5? What do 1, 3, 7, 15, 31 and 63 have in common that will enable your class to guess a formula?

It would be easier if the numbers were a little bigger; say 2, 4, 8, 16, 32 and 64. These are again powers of 2: specifically 2^1, 2^2, 2^3, 2^4, 2^5, and 2^6. Here the powers exactly reflect the number of discs. So it is worth making the next conjecture.

Conjecture 3: The number of moves for d discs is $2^d - 1$.

Don't worry if some students find this hard to follow; we will return to this in Level 3.

We now have a possible formula and it should be easier to work out the number of moves for 64 discs than it was using either Conjecture 1 or Conjecture 2. But you might need a calculator that has a large number of decimal places to get an exact value of $2^{63} - 1$.

(Don't forget to divide by 365 to get the number of years that the monks will take to move all of their discs, assuming they don't make any errors.)

Where to from here?

I   Get your class to think about what *could* be done at this point and what *should* be done (even if they are unable to do it).

Level 3: Proofs

Problem

Can you prove any of these conjectures about how many steps are required?

Problem steps

Step 1

In this level we want to show that the following three conjectures are true.

> **Conjecture 1:** The number of moves for d discs is twice the number of moves for $(d - 1)$ discs plus 1.
>
> **Conjecture 3:** The number of moves for d discs is $2^d - 1$.
>
> **Conjecture 2:** The number of moves for d discs is 2^{d-1} more than the number of moves for $(d - 1)$ discs.

We have presented the conjectures in a different order here, because they are easier to prove this way. Concentrate on Conjecture 1 first, then use that to settle Conjecture 3; from there, Conjecture 2 isn't too hard. It is probably a good idea not to use algebra at first. Go through the arguments below, starting with small values of d (even 1), then building up until students see that the conjectures are true for specific values. At that point, bring in the general terms and use algebra to complete the deal.

Step 2

Look for ideas around the movement of the biggest disc. Remember that we need to get the pile of discs from the left-most peg to the right-most.

The situation in Figure 1.12 is crucial. At some point we want to move the biggest disc from the left peg to the right. This can only happen if *all* the other discs are on the middle peg, and this has to have been done in the *smallest* number of moves.

Figure 1.12: What is going on when the biggest disc is about to move?

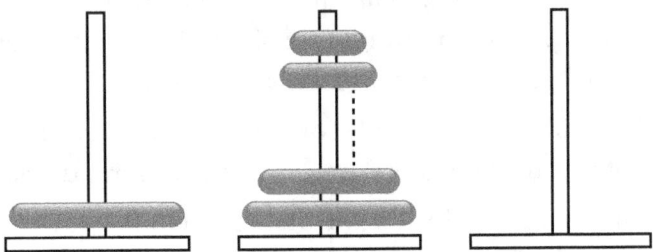

Then we can move the biggest disc, and after that the middle stack of discs.

Adding everything up we get:

The number of moves needed to move d discs

$\quad$ = (number of moves that it takes to move $d - 1$ discs) + 1
$\quad$ + (number of moves that it takes to move $d - 1$ discs)
$\quad$ = 2(number of moves that it takes to move $d - 1$ discs) + 1

Students may understand this better if you first put in $d = 1$, then $d = 2$ and so on until they have the idea. Then you can write down the general case on page 37, and so prove Conjecture 1.

Step 3

What about Conjecture 3?

One way to do this involves working ever upwards:

- 1 disc: we know this takes 1 move and $1 = 2^1 - 1$
- 2 discs: two discs takes 2(the number of moves for 1 disc) + 1 = $2(2^1 - 1) + 1 = 2^2 - 1$
- 3 discs: three discs takes 2(the number of moves for 2 discs) + 1 = $2(2^2 - 1) + 1 = 2^3 - 1$
- and so on, repeating forever.

We can formalise this in the following way:

Clearly the number of moves for 1 disc is $2^1 - 1$.

Now suppose that the number of moves for k discs is $2^k - 1$.

From the proved Conjecture 1, we know that the number of moves for $k + 1$ discs is 2(the number of moves that it takes to move k discs) + 1 = $2(2^k - 2) + 1 = 2^{k+1} - 1$

So the pattern is preserved at every step.

To demonstrate this for students, you could go through the stages like this:

- We know that the 'suppose' is true for $k = 1$, so it has to be true for $k = 2$.
- Now we know that the 'suppose' is true for $k = 2$, so it has to be true for $k = 3$.
- Now we know that the 'suppose' is true for $k = 3$, so it has to be true for $k = 4$.
- Now we know that the 'suppose' is true for $k = 4$, so it has to be true for $k = 5$.
- Now we know that the 'suppose' is true for $k = 5$, so it has to be true for $k = 6$.

And so it goes on forever, and for every number of discs d. So Conjecture 3 is true.

This is essentially a proof by *mathematical induction*. Such proofs are valuable where something needs to be proved for an infinite set of numbers (in this case d, the number of discs). In general, mathematical induction follows this pattern:

Aim: We are looking to prove a formula for all values of a certain thing. (In the *Tower of Hanoi* problem it was to show there are $2^d - 1$ moves for d discs.)

First: Show that the formula holds for the first case. (In the *Tower of Hanoi* problem we showed that when $d = 1$, there were $2^d - 1 = 2^1 - 1 = 1$ moves needed.)

Second: Show that if the formula holds for one case, it holds for the next. (In the *Tower of Hanoi* problem, we showed that if there were $2^{d-1} - 1$ moves for $d - 1$ discs, we get $2^d - 1$ moves for d discs.)

This proof technique is used a lot in arguments that require something to be proved for all or most natural numbers. It is most commonly used in mathematical areas such as number theory and combinatorics, where natural numbers are commonly involved.

One final comment: in the process of finding the number of moves, we have also shown by induction that it is possible to move any number of discs from the left to the right peg. We just start with two discs and use the ideas around Figure 1.10 to move more and more discs. In the next level we give a more formal algorithm for moving the discs.

Step 4

Now ask students to work out how many moves are needed to take 64 discs from the left peg and move them to the right peg. It is easy to say that the number of moves is $2^{64} - 1$, but how do you work this out?

You will need something that can handle 20 digits; a simple calculator won't be able to cope, but CAS calculators, online calculators and/or the calculator function on a computer should be able to do this. The answer that you get is 18 446 744 073 709 551 615.

In scientific notation this is about 1.84×10^{19}. To find the number of years the monks will need to move all the discs, divide this number by 365. Whatever the result is, students can be assured that this won't happen in their lifetime. In fact, how does this compare with the number of atoms in the universe or even the age of the universe?

Step 5

Why do successive numbers of moves differ by increasing powers of 2 (Conjecture 2)? Encourage students to use any information they currently have and work with some specific examples.

Perhaps the easiest way to tackle this is to look at the result of the proved Conjecture 3. Since the number of moves for four discs is $2^4 - 1$, and the number of moves for five discs is $2^5 - 1$, then the difference between these two is $(2^5 - 1) - (2^4 - 1) = 2^5 - 2^4$.

How can we simplify this? There is a common factor of 2, so now $2^5 - 2^4 = 2^4 (2 - 1) = 2^4$. And that should work in general: $2^d - 2^{d-1} = 2^{d-1} (2 - 1) = 2^{d-1}$.

Where to from here?

▌ What ideas do the students have for progress here? What questions have been left open?

▌ We move on to working out a method (algorithm) for moving the discs in the next level.

Level 4: How to move the discs

Problem
Could you program a computer to move the discs?

Problem steps

Step 1
In this level we will work on an algorithm to move any number of discs.

Ask your students: what would you tell a friend who wanted to move 15 discs? How should your friend move the next disc, whatever that disc should be? What have we learned already that might help us out here?

It is always good to experiment to get ideas. If students can go straight to an algorithm for moving any number of discs, good on them! It is more likely that they will have to look at the specific examples in Levels 1 and 2. What is going on there? What happens to the first disc, the second disc and so on?

Step 2
What does happen to the first, second, third and fourth disc each time? Do the moves of these discs vary at all?

Table 1.6 lays out the first moves of the first four discs so that we can look for patterns. In the table, *R* stands for the right peg and *M* for the middle peg; a dash means the given disc doesn't exist.

Table 1.6: The initial moves of the first four discs

	1 disc	2 discs	3 discs	4 discs	5 discs
1st disc	R	M	R	M	R
2nd disc	--	R	M	R	M
3rd disc	--	--	R	M	R
4th disc	--	--	--	R	M

From the data so far, we can make these statements.

- An odd-numbered disc moves to the right peg if there is an odd number of discs to start with, or to the middle if there is an even number of discs.
- An even-numbered disc moves to the right peg if there is an even number of discs to start with, or to the middle if there is an odd number of discs.

Step 3
What happens before the third disc is able to move in the three-disc problem? What happens before the fourth disc can move in the four-disc problem?

The thing to note here is that the movement of *all* of the discs from the left peg to the right is achieved in a series of moves of 1, 2 or 3 discs. Before you can move the third disc, you have to have made a pile of two discs on one peg. The reason for this is that to move disc 3 from the pile of discs on the left peg, it has to go to a 'free' peg—you can't put the third disc on top of discs 1 or 2. That forces discs 1 and 2 to have already formed a pile on one of the other pegs. This is exactly the same as the argument for the proof of Conjecture 1 in Step 3 of the last level.

The same thing happens with the fourth disc on the left peg. There has to be a free peg to which the disc can be moved, which is only possible if the first three discs already form a legitimate pile on some peg. To produce the smallest number of moves, this peg should be the middle one.

Step 4

Does that help you to produce an algorithm to move six discs? How would you tell a friend to move those discs?

Moving a pile of d discs is done through a series of moves that produce first a pile with one disc, then a pile of two discs, and so on up to the final d discs. The only limitation is that disc 1 first moves to the right peg if d is odd, and to the middle peg if d is even. (That is easy to remember if we think about what happens if there is only one disc.)

But if your 'friend' happens to be somewhere in the middle of moving 64 blocks, it might be difficult to know exactly what to do next. How do we know precisely what should be the next move?

Step 5

From here on we write numbers in place of the discs with the smallest number being 1, the next smallest 2, and so on up to the largest disc.

What are the next four moves in Figure 1.13? How are we going to move disc 4, and to what peg does it need to go?

The aim here is to get discs 1 to 4 on the middle peg. So you need to have disc 1 on 2 on 3 in the middle.

Figure 1.13: Part-way through moving five discs

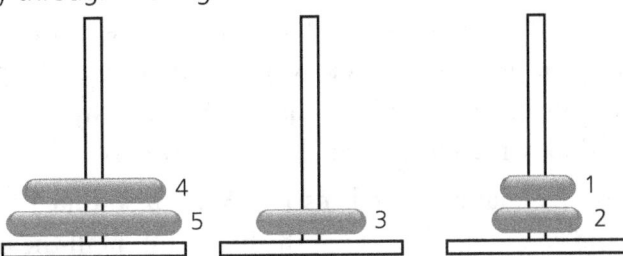

From Figure 1.13, we can see that 1 goes on top of 4; 2 on top of 3; 1 goes on top of 2; 4 goes to the right peg.

Encourage students to go through to completion here, and to keep note of these moves in an attempt to find a pattern.

Step 6

What are the next seven moves in Figure 1.14?

Figure 1.14: Part-way through moving nine discs

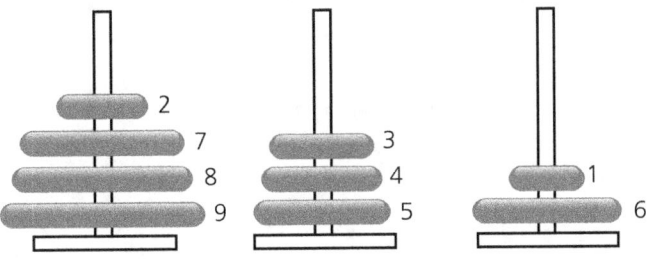

Here we see that 1 goes on 2; 3 goes on 6; 1 goes on 4; 2 goes on 3; 4 goes on 7; 1 goes on 4.

Again, encourage students to complete the work and to make notes so as to help find a pattern. Look carefully to see where the odd discs go. Where do the even discs go?

Step 7

In general, your students know where to move disc 1 from Step 2. How do you know where to move *any given disc* at any time in the process of moving three, six or eight discs? What patterns can you see from the previous two steps?

Students may be able to see the following rules from what has been done so far.

1. Move disc 1 to the right peg if the number of discs is odd, and to the left peg if the number of discs is even.
2. Any time there is a single disc on a peg, move it to the free peg—the peg with no discs on it.
3. At any other time, don't undo the previous step, but move the smallest disc on top of an even disc if it is odd and to an odd disc if it is even.

The main part of the third rule can be summarised by saying that you can't place any odd disc on an odd disc, nor any even disc on an even disc.

Where to from here?

- Can students justify the algorithm? Why do you not want to put an odd disc on another odd disc?
- Is it true that the biggest disc moves once, the next-biggest twice, the next-next-biggest four times and so on? Does this explain why the number of moves is $1 + 2 + 2^2 + \ldots$?
- What if the shrine had *four* pegs? Now we can move discs from the left peg to the right peg more easily because we can just forget about one of the pegs. What is the smallest number of moves required for d discs? (As far as we know, nobody knows the answer to this question, which is known as Reve's puzzle.)
- Another variant problem: label the d discs 1, 2, 3, ..., d so that 1 is the smallest, 2 is the next smallest and so on, up to d being the biggest. Put all the odd numbered discs, in order from smallest to largest, on the left peg and the even ones in order on the right peg. The object is to move the odd numbered discs to the right peg and the even numbered discs to the left peg. The usual rule applies: no bigger disc can be put on top of a smaller disc. What is the smallest number of moves required?
- There are many other variants and special cases for this concept, such as the situation where there are twice as many pegs as discs. You may like to look these up yourself, or have students research them online.

- An extension activity, 'The Caliph's problem', can be downloaded from the series website.

42   Creative Activities in Mathematics: Book 3

CHAPTER 3: NIM-LIKE GAMES

Initial problem

There are 21 tokens on the table. Robyn and Kuparr play alternately by taking one or two tokens from the table. The player who takes the last token wins; if there are two tokens left, then a player can remove both of them to win. Both players are trying to win.

If Robyn plays first, is she bound to win?

Background information

This problem is a variation and extension of the '12 game' activity in *Creative Activities in Mathematics Book 1*. This game is like the game of Nim, which is introduced at Level 4. Nim has a long history, and may possibly have begun in ancient China; whether or not this is so, the game didn't reach Europe until the 1500s. Harvard mathematician C L Bouton gave it the name *Nim*, and developed the full solution of the game in 1901. Nim has many variations, and variables can include how many tokens can be taken from the pile at one time, how many tokens are in the pile and how many piles there to start with.

In Level 1 we consider the 21 game and encourage students to play it and to consider who will win and why. Level 1 should be accessible to most students from Year 6 and up.

The two-pile version of the game is treated in Level 2 and the three-pile version in Level 3. These two levels give students practise in looking at situations, generalising them and trying to prove their conjectures. Level 2 is appropriate for many Year 7 students and most Year 8 students, while Level 3 is suitable for most Year 8 students and some advanced Year 7 students.

Finally, Level 4 moves on to the Nim game proper; it is appropriate for some Year 8 and most Year 9 students.

Some of the details in the various levels are quite difficult to understand. You will need to judge how far your students can go. But all students can get something out of this problem, so take them as far as you think they can go.

Table 1.7: Australian Curriculum content descriptions for the *Nim-like games* activity

Activity level	Problem	Content descriptions
1	One-pile 21 game	*Year 6* Continue and create sequences involving whole numbers, fractions and decimals. Describe the rule used to create the sequence (ACMNA133) *Year 7* Introduce the concept of variables as a way of representing numbers using letters (ACMNA175)
2	Two-pile 21 game	*Year 7* ACMNA175 (see above)
3	Three-pile 21 game	*Year 7* ACMNA175 (see above)
4	Nim	*Year 7* ACMNA175 (see above)

Note: Although Levels 2, 3 and 4 only require mathematics content knowledge at Year 7 level, they require mathematical maturity and sophistication above that age level.

Big ideas

» How to use divisibility by 3
» How both persons in a two-player game might be able to win

Problem aims

» To see patterns in a game and to be able to justify these patterns

Key concepts

» That a number is divisible by 3 if the sum of its digits is divisible by 3

Possible heuristics/strategies

» Playing the game
» Using simpler cases
» Being systematic
» Making a table

Resources

» Blocks, chips or similar items to be used as tokens

Special notes

» It is useful to know that $3k$ is any multiple of 3; $3k + 1$ is a multiple of 3 plus 1; and $3k + 2$ is a multiple of 3 plus 2. However, the algebraic method we present in Level 1 can be avoided.

» *If and only if* is a short way to represent two implications. We know that if a number is divisible by 3 then the sum of its digits is divisible by 3. Similarly, if the sum of the digits of a number is divisible by 3, the number is divisible by 3. We can reduce this by putting the two implications together in 'A number is divisible by 3 *if and only if* the sum of its digits is divisible by 3'. In general, if A implies B and B implies A, we can say that A is true *if and only if* B is also true.

» Binary numbers are only needed if a student is doing the investigation part of Level 4. These are numbers in base 2. For example, the binary number 101011 is $1 \times 2^5 + 0 \times 2^4 + 1 \times 2^3 + 0 \times 2^2 + 1 \times 2^1 + 1 \times 2^0 = 43$.

Level 1: One-pile 21

Problem

There are 21 tokens on the table. Robyn and Kuparr play alternately by taking one or two tokens from the table. The player who takes the last token wins; if there are two tokens left, then a player can remove both of them to win. Both players are trying to win.

If Robyn plays first, is she bound to win?

Problem steps

Step 1

(Note that Robyn always goes first, in this and in the following levels.)

The difficult part about this game is that students may take a long time to realise that there is a winning strategy here and that by playing a certain way, either the first or second player must always win. Let the class play in pairs for a while, then ask them to vote on the following three options.

1. The first player will always win.
2. The second player will always win.
3. It is just a matter of luck and anyone can win.

Whichever one they choose, ask them to justify their choice.

The correct answer is the second player should always win. If they haven't seen this, they should try small cases. Let them do the experiments and complete Table 1.8.

Table 1.8: The winners for small numbers of tokens

Number of tokens	1	2	3	4	5	6	7	8	9	10	11	12	13	14	15	16
Who wins	R	R	K	R	R	K										

What does Table 1.8 suggest are the best numbers for Robyn? Which are the best for Kuparr?

Now repeat the vote with the three options: the first player will always win; the second player will always win; anyone can win. Can they justify their choices?

Kuparr can always win the 21 game by watching what Robyn does. If she removes one token he should remove two; if she removes two tokens he should remove one. This way he forces Robyn to be faced in turn with 18, 15, 12, 9, 6, and then three tokens. Kuparr clearly wins when there are only three tokens.

Step 2

We've seen that Kuparr will win when playing second when there are 21 tokens. For what other numbers of tokens will Kuparr win as second player?

Let students experiment and conjecture until they see that any number of tokens that is a multiple of 3 will lead to a Kuparr win. He can use exactly the same strategy that he did on the 21 game and, on his turn, reduce the pile to a smaller multiple of 3.

Step 3

Suppose that there were 43 tokens. Who would win and why?

Let the students find out for themselves that Robyn will win here by removing one token on her first play. Let them also explain to you *why* she will win. This is because 43 − 1 = 42, which is a multiple of 3. We have seen that the second player to tackle a pile that is a multiple of 3 will always win. By removing one token from the 43 tokens, Robyn has put herself in the position of second player to a pile that is now a multiple of 3.

Step 4

Suppose that there were 38 tokens. Who would win and why?

Let the students find out for themselves that Robin will win here by removing two tokens on her first play. Let them also explain to you why she will win. This is because 38 − 2 = 36 which is a multiple of 3. As in Step 3, this makes her the second player to a pile that is now a multiple of 3, from which point she will always win.

Step 5

Your class is now in a position to *generalise* this problem. This means that they can tell, from the start, who will win with *any* given number of tokens. Discuss this with the class and see what they come up with.

Suppose that there were s tokens. Who would win and why?

Let the students find out for themselves that:

- if $s = 3k$, where k is any whole number, then Kuparr will win by using the same strategy that he did with $s = 21$
- if $s = 3k + 1$, where k is any whole number, then Robyn will win here by removing one token on her first play
- if $s = 3k + 2$, where k is any whole number, then Robyn will win here by removing two token on her first play.

If your students aren't up to algebra yet, use a very large number here, like 5329. If they can see how to tackle this problem, then they intuitively know how to do it for any number. If they have trouble with $3k$, $3k + 1$ and $3k + 2$, ask them what happens for 5328, 5329 and 5330.

We don't expect that all of your students will be comfortable with this algebraic way of presenting the generalisation. So long as they have the idea and could tell you who will win and how with any number of tokens you give them, that is enough.

Where to from here?

- How will things change if Robyn and Kuparr in turn are able to take one, two or three tokens from the pile? If they both play as well as possible, who will win this game? Ask the students how they plan to tackle this problem. Do they know the answer before they start? Students should find that Kuparr will win if and only if the number of tokens is a multiple of 4. Extend further to any number of tokens on the pile.

- What happens if 1, 2, 3, ... *n tokens* can be removed at a time?

Level 2: Two-pile 21

Problem

Suppose that there are two piles of tokens with 15 in one pile and 6 in the other. Robyn and Kuparr can take either one or two tokens from *one* of the piles when it is their turn. Both of them are playing as well as possible.

Can either Robyn or Kuparr find a strategy to make sure that they always win?

Problem steps

Step 1

(It is important that students play the game from Level 1 before they start on this problem.)

Let the students play the game and come up with a strategy for one or the other player to win.

They should find that Kuparr will win by using his strategy from the one-pile situation in Level 1. After Robyn's move, he makes sure that the pile she took tokens from is reduced by 3. Eventually one of the piles will be reduced to 0, while the other one is a multiple of 3. Robyn will need to take the first token(s) from this pile and so Kuparr will win in the usual way.

Step 2

What if there are two piles consisting of 10 and 11 tokens? Who will win here?

This game is more complicated so make sure your students have made a good attempt at analysing it by playing several games before they make any conjectures. There are lots of possible moves and it may be harder to see what is happening here.

Try to get your students to come up with the suggestions of:

- playing some simpler games first to try to develop some intuition
- being systematic in the numbers of tokens they use
- making a table to record their results.

See if they can find their way to the answer, which is that Robyn will win.

Step 3

What about one pile with one token and the other with one, two, three or even t tokens? What about one pile with two tokens and the other with two, three or even t tokens? What about more complicated cases?

We will use the notation (s, t) to denote s tokens in pile 1 and t tokens in pile 2. (Note that there is no difference between an (s, t) game and a (t, s) game.) We show the results of some small games in Table 1.9.

Table 1.9: Small games with two piles

Pile 1	Pile 2	Who wins?	Why?
1	1	Kuparr	Robyn takes one and Kuparr takes the other
1	2	Robyn	Robyn takes one from pile 2 and reduces the game to a (1, 1) game. She wins as she is the second player to the (1, 1) game
1	3	Robyn	Robyn takes two from pile 2 and reduces the game to a (1, 1) game
1	t	Kuparr if $t = 3k + 1$	If Robyn takes the single token from pile 1, Kuparr takes one from pile 2. Otherwise, Kuparr makes sure that pile 2 goes down by three and waits till either Robyn takes the token from pile 1 or the game is reduced to the (1, 1) game
		Robyn if $t = 3k$ or $3k + 2$	Robyn reduces the game to a (1, $3k + 1$) game
2	2	Kuparr	Kuparr takes the same number of tokens from pile 2 that Robyn did from pile 1
2	3	Robyn	Robyn reduces the game to a (2, 2) game
2	t	Kuparr if $t = 3k + 2$	Kuparr repeats what Robyn does but on the other pile
		Robyn if $t = 3k$ or $3k + 1$	Robyn reduces the game to either a (2, $3k + 2$) game or a (1, $3k + 1$) game
3	3	Kuparr	Kuparr repeats what Robyn does but on the other pile
3	t	Kuparr if $t = 3k$	Kuparr repeats what Robyn does but on the other pile
		Robyn if $t = 3k + 1$ or $3k + 2$	Robyn reduces the game to either a (1, $3k + 1$) game or a (2, $3k + 2$) game

Give students time to think about this and scaffold them if or when they need it. This scaffolding may be just encouragement ('you are pretty close here', 'that is good, but what if …?', 'have you thought of …?'), or it may be helping them over an error or suggesting some new, small, specific case to be tackled.

Step 4

So who wins the (s, t) game?

Here s and t can be anything you like. Give this to them as big numbers first, say the (98, 105) game. Why do they think that this one is a win for Robyn or Kuparr? How will Robyn or Kuparr win it?

When they have mastered *any* two big numbers, go back to s and t and ask them to think up a conjecture. We suggest the following:

> **Conjecture:** Kuparr will win if s and t have the same remainder when divided by 3. Robyn will win otherwise.

(If they know the concept of equivalence modulo 3, then they can write $s \equiv t \pmod 3$. You may even want to introduce them to this notation.)

This conjecture is equivalent to the one below:

> **Conjecture:** Kuparr will win *if and only if* s and t have the same remainder when divided by 3.

The proof of these conjectures follows from Table 1.9. If s and t have the same remainder when divided by 3, then whatever Robyn moves from pile 1 or pile 2, Kuparr can adjust the numbers so that it is now an (s', t') game with s' and t' having the same remainders when divided by 3.

As the number of tokens gets smaller, someone must take the last token from a pile. If Robyn does this, Kuparr must be able to reduce the other pile to a multiple of 3 and then he wins. If Kuparr removes the last token from a pile, then the other pile has to have a multiple of 3 in it, and again Kuparr wins.

If s and t don't have the same remainder when divided by 3, then Robyn can take one or two tokens away to make it an (s', t') game with s' and t' having the same remainder when divided by 3. Kuparr is now the first player to meet this situation; he is in the same position as Robyn was in the last paragraph, so he loses.

Where to from here?

- There are all sorts of directions that you might go from here. Listen to what your students suggest. Something that might be interesting is to determine the probability of Kuparr winning the one- and two-pile games, if he and Robyn are playing as well as possible.
- It is also possible to look at removing a different number of tokens on each turn.

Level 3: Three-pile 21

Problem

Now Robyn and Kuparr are faced with three piles of tokens. These piles have three, four and five tokens in them respectively. They can each take one or two tokens from one of the piles when it is their turn.

Who will win this game if Robyn and Kuparr each play as well as they can?

Problem steps

Step 1

Some parts of this problem rely on material from Levels 1 and 2, so it is worth displaying any notes or tables from these prominently around the room.

Emphasise that Robyn is the first player and Kuparr the second player. So either player coming at a given game as first player will win games that Robyn won in Level 1 or Level 2, even if that first player is Kuparr this time. The same thing is true for the second player.

As your students have tackled the earlier levels, they may have a feel for what is going on here. This may lead them to make a conjecture right at the start, but they have to be able to *prove* this conjecture. So how can they do this?

It wouldn't hurt for them to systematically do some smaller cases; this may enhance their intuition and help see how to work things out. Have them start by drawing up a table of small values and including who wins the games and how. The table should be something like Table 1.9, with an extra column for pile 3. What results does this table suggest?

Step 2

Give different small games to different pairs of students to work on. The class can then pool their results. Allow for more than one pair to have specific numbers of tokens. This will provide a check for the students' work.

- In the (1, 1, 1) game Robyn must win.
- In the (1, 1, 2) game Robyn must win by taking both tokens from pile 3.
- In the (1, 1, 3) game Kuparr must win. If Robyn tackles pile 3, Kuparr reduces it to 0 and then Robyn is faced with the (1, 1) game, which she loses. If she takes a token from pile 1 or 2, Kuparr will take the other single pile. This leaves Robyn with a pile divisible by 3, which she will lose.
- In the (1, 1, u) game, Kuparr will win if $u = 3k$ and lose otherwise. It all depends on how Robyn reduces the game. She should get a clue from the three games above.

Step 3

What about (1, 2, u) games?

- In the (1, 2, 2) game Robyn must win by taking the token from pile 1. This reduces the game to a (2, 2) game that the second player will win. Robyn is the second player to get to the (2, 2) game.

- In the (1, 2, 3) game Robyn must win by taking one token from pile 2. This is now a (1, 1, 3) game, which the second player (now Robyn) will win.
- In the (1, 2, 4) game Robyn must win by taking the two tokens away from pile 2. This is now a (1, 4) game, which the second player will win. In this case, Robyn is the second player to this game.

Does that mean that Robyn will always win the (1, 2, u) game?

Yes, it does. That is a bit of a surprise! We'll do each of the cases for u separately.

- If $u = 3k$, then Robyn removes one of the two tokens on pile 2 to turn it into a (1, 1, 3k) game which is a win for the second player: Robyn.
- If $u = 3k + 1$, then Robyn takes two tokens from pile 2 and wins the ensuing (1, 3k + 1) game.
- If $u = 3k + 2$, then Robyn takes the one token from pile 1 and wins the (2, 3k + 2) game.

Step 4

So what happens with the (1, 3, u), the (2, 2, u), the (2, 3, u) and so on games? Can your students make a conjecture out of all of this?

Let all conjectures be considered and discussed. Does anyone know a counter-example? Does the conjecture fit all of the examples in the table?

Once they are satisfied with a class conjecture, get some students to try to prove the conjecture and others to try to find a counter-example. If there is a counter-example, the conjecture needs to be reconsidered.

We suggest the following:

> **Conjecture:** Kuparr will win (s, t, u) games where one of s, t and u is divisible by 3 and the other two have the same remainder when divided by 3. Robyn will win all of the others.

Kuparr's tactic here is to try to keep the condition 'one of s, t and u is divisible by 3 and the other two have the same remainder when divided by 3' intact after Robyn plays.

Suppose that s and t have the same reminder and u is divisible by 3. It Robyn takes something off pile 1, then Kuparr takes the same off pile 2. If Robyn takes something off pile 3, then Kuparr makes sure that that pile still has a multiple of 3 left in it. Let the students play with this idea until they are convinced that it works for Kuparr.

On the other hand, if the condition 'one of s, t and u is divisible by 3 and the other two have the same remainder when divided by 3' doesn't hold, then Robyn should make sure that it does straight away. Let your students convince themselves that this can be done.

Now get students to play against each other with some large numbers to make sure that they understand what is going on.

Where to from here?

- Students might look at removing one, two or three tokens at a time, or even 1, 2, 3, ..., n. Then they might have a look at problems using four, five or more tokens.
- On a completely different tack, you might ask: What is the probability that Kuparr will win a game chosen at random, given that he and Robyn can play as well as anyone else in the world and want to win?

Level 4: Nim

Problem

Kuparr and Robyn are facing two piles of tokens. As usual, Robyn goes first.

But the rules are slightly different. This time she can take as many tokens at a time as she likes, provided they are *only* from one pile. To make things fair, Kuparr has the same choice of tokens.

If both players play as well as possible, who will win?

Problem steps

Step 1

This may not seem like a fair question, since we haven't told you how many tokens there are in each pile. So get your students to play in pairs and choose their own sized piles. What do they find?

This is repeating the kind of approach that was taken in the previous levels, but now your students are in charge.

Step 2

Clearly we are looking for some conjecture that we might prove.

Suppose that there are s tokens on pile 1 and t tokens on pile 2. When will Robyn win?

Suppose first of all that $s = t$. This is a win for Kuparr; whatever Robyn does to one pile, he does to the other.

Step 3

What if $s \neq t$?

If $s \neq t$ then Robyn wins. On her move, she just makes sure that the two piles are equal. Then she uses Kuparr's strategy from Step 2.

Step 4

Can your students write up a proof of this? We've already done the skeleton of a proof, but can they make it watertight and defend it against all challenges? That is, can they prove it?

We'll write it up as a theorem that we then prove.

Theorem: Kuparr wins if $s = t$ and Robyn wins if $s \neq t$.

Proof:

Case 1: $s = t$. Robyn has to take some number r from one of the piles. So the (s, s) game becomes an $(s - r, s)$ game. Suppose that Kuparr now takes r from the other pile to produce an $(s - r, s - r)$ game. If $s = r$, then the game is finished and Kuparr wins. Otherwise Robyn will take some more tokens from one of the piles. But Kuparr repeats his strategy and the piles are now level again. Because the number of tokens in each pile is decreasing, after playing like this for a number of steps, Robyn has to take the last token in one of the piles. Kuparr then takes the last token from the other pile and wins.

Case 2: s < t. Robyn plays first and takes *t – s* tokens from the bigger pile. Kuparr is now in Robyn's position from Case 1 and now Robyn plays the way that Kuparr did there.

It is clear that either the two piles have an equal number of tokens or they don't. So the two cases in the theorem cover all eventualities. We thus have a result that follows from the theorem; such results are called *corollaries*.

Corollary: Kuparr wins *if and only if s = t*.

Proof: If *s = t*, then Kuparr wins from the theorem. So Kuparr wins if *s = t*.

If *s ≠ t*, then Robyn wins. So Kuparr does not win if *s ≠ t*.

As a result, Kuparr wins if *s = t* and not otherwise. So Kuparr wins *if and only if s = t*.

Step 5

What happens if there are *three* piles?

Let the students choose the numbers in each pile and, as they come up with answers, put them on the board for other students to check. What conjectures can they come up with?

It should be possible to see that if all piles have the same number of tokens, then Robyn will win; she just takes all of one pile and Kuparr is faced with the (*s*, *s*) game and will lose. Similarly, if two piles are the same and the third different, Robyn wins by reducing the third pile to zero.

But what if all three piles are different?

First, take the simplest case of (1, 2, 3). Even this may take some discussion to resolve in Kuparr's favour. Consider all of Robyn's options and show that whatever she does, Kuparr can win. Let your students play with other games to see if they can find a pattern.

Step 6

It is likely that not very much progress will be made. There are at least two ways to proceed. You might hint that it has something to do with numbers in base 2, or you might like to set it as a online research exercise (links can be found on the series website).

In the latter case, suggest that students talk about it in class and show how that method applies to particular small cases such as (2, 3, 5) and (2, 4, 5). They might discuss the general idea and explain why it works. After they have a full understanding of the method, they can write up an explanation of the winning strategy, both for the two-pile and three-pile games.

Where to from here?

- Students might learn quite a bit by delving into the history of Nim and its variations and understanding the theory of these games. This might include 'nimbers' and 'nimsums'. This exercise will give them practise in reading new maths, understanding it and writing it up for others to understand.
- Think about how the online method works for four-pile games and more.

PART 2: MEASUREMENT AND GEOMETRY

Part 2 presents three activities centred on the Number and Algebra strand.

Table 2.1: Measurement and Geometry activities

Problem	Big ideas
Hidden treasure	- Ruler and compass constructions - Basic ideas linking circles and simple polygons
Tessellations	- Exploring shapes and how they fit together—the sum of the angles at a point is 360° - The sum of the interior angles of polygons
How high is a building?	- Measuring lengths and angles - Using ratio and trigonometry

Some reminders before you use these tasks in your classroom:

1. The questions in the text are ones you can ask your students. You are likely to be able to produce similar, more immediately relevant ones for your particular students as you work on these activities with them.
2. We have suggested links to the Years in the Australian Curriculum: Mathematics for all the Levels in each activity but, given that there will be a spread of ability in your classes, you should take these as a guide only. Take the opportunity to encourage every student to the edge of their comfort zone.
3. To take all students further, sometimes you can omit some of the later steps of a Level in favour of the early steps in the following Level.

CHAPTER 4:
HIDDEN TREASURE

Initial problem

An ancient treasure has been hidden in a graveyard an equal distance from each of four headstones. These headstones are on the corners of a square. Can you find the treasure?

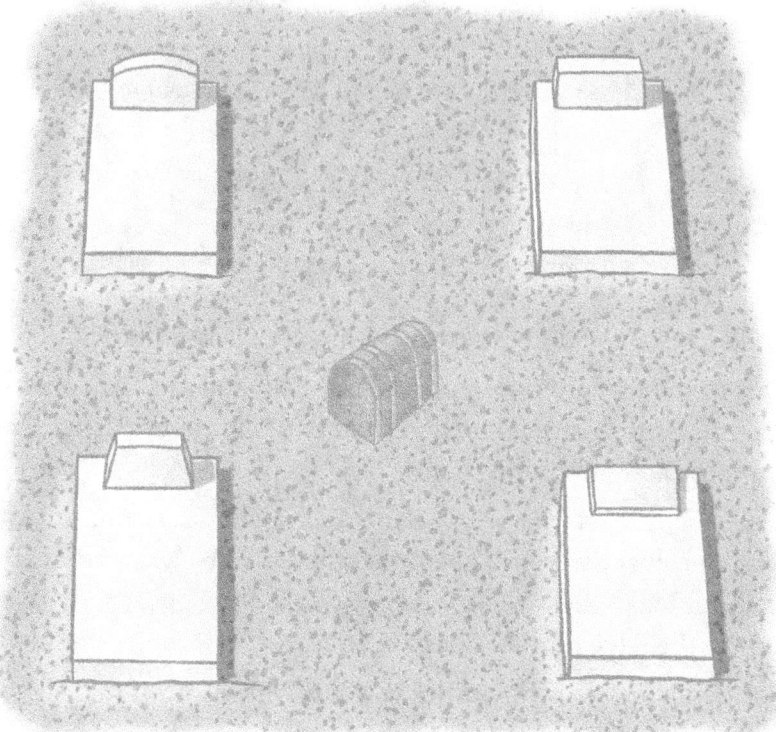

Background information

This activity is based on material whose origins can be found in Euclid's famous book *The Elements*. This book served as a text on geometry for more than 2000 years—quite an achievement! Euclid, on the other hand, is a mysterious figure about whom little is known. He may or may not have existed; he may have been the leader of a group of mathematicians, or the mathematicians may have invented him as a group alias as they wrote *The Elements*. Whatever the situation, *The Elements* contains all of the mathematics that was known at the time it was written; some of the contents were probably Euclid's own, but most of it is a compilation of the findings of others.

 Both Euclid and *The Elements* are interesting subjects to research, and it is worth exploring online to find out more than we can include here. The series website contains a link to the Story of Mathematics site, which is worth exploring for many points of history of mathematics. At the least, we suggest that you give it to your students to look up the 'List of important mathematicians'.

This activity is largely about ruler and compass constructions, but a great deal of progress can be made without them. If possible, it is worth doing much of the actual construction work outside in the schoolyard or playground.

In Level 1 we investigate finding hidden treasure using the vertices of a square. This involves constructing squares and their diagonals. The part of Level 1 that doesn't involve ruler and compass constructions and proofs is accessible to Year 7 students. With help, most Year 8 students can do all aspects of this level.

Level 2 extends the concept to triangles. Here the construction of the triangle is straightforward but finding the treasure is relatively hard. Level 2 will be accessible to some Year 7 students and most Year 8 students, except perhaps for the proof aspect, and it should be suitable for all Year 9 students.

Level 3 looks at the same scenario but with quadrilaterals. Some of Level 3 can be tackled by advanced Year 7 and Year 8 students, but this is mainly an activity for Year 9 students.

Level 4 moves on to hexagons, and is for Year 9 and Year 10 students.

It's worth noting that while Levels 3 and 4 are primarily for Year 9 and 10 students, they are fundamentally explorations and extensions of Year 8 content descriptions from the curriculum. The need to take these concepts in new directions and apply them in new circumstances is what makes these activities more suitable for older students.

Table 2.2: Australian Curriculum content descriptions for the *Hidden treasure* activity

Activity level	Problem	Content descriptions
1	Four pegs	*Year 7* Classify triangles according to their side and angle properties and describe quadrilaterals (ACMMG165) *Year 8* Investigate the relationship between features of circles such as circumference, area, radius and diameter. Use formulas to solve problems involving circumference and area (ACMMG197) Develop the conditions for congruence of triangles (ACMMG201)
2	Three pegs	*Year 7* ACMMG165 (see above) *Year 8* ACMMG197 (see above) Define congruence of plane shapes using transformations (ACMMG200) ACMMG201 (see above)
3	Any four points	*Year 8* ACMMG200 (see above) ACMMG201 (see above)
4	Regular hexagons	*Year 8* ACMMG197 (see above)

Proofs are one of the most important elements of geometry, but also one of the most difficult for many students—and for more than a few teachers. If you find the proofs in this activity complex, reread them several times and get comfortable with them before putting the activity into practise with your class.

Big ideas
» Ruler and compass constructions
» Basic ideas linking circles and simple polygons

Problem aims
» Increasing students' geometric intuition, especially of circles and related quadrilaterals
» Practising geometric constructions

Key concepts
» Similarity of triangles
» Names and properties of quadrilaterals
» Pythagoras' theorem

Possible heuristics/strategies
» Trial and error

Concrete materials
» String or rope
» Tools to measure distances longer than 3 metres
» Pegs
» Protractor
» Compass
» Ruler

Level 1: Four pegs

Problem

An ancient treasure has been hidden in a graveyard an equal distance from each of four headstones. These headstones are on the corners of a square. Can you find the treasure?

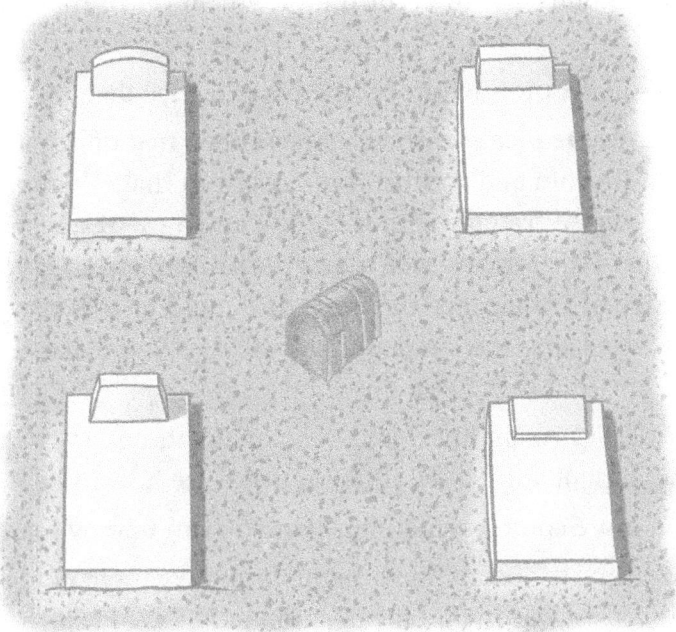

Problem steps

Step 1

Discuss with the class how they would go about finding the treasure. The simplest way is to join pairs of diagonally opposite headstones with string. Where the two strings intersect, the students can expectantly dig for the treasure.

But why does this work? Discuss this with your students.

Also ask them if they can think of any other way to find the treasure. For example, they might measure the distance between two diagonally opposite headstones, then measure to the midpoint between these two headstones.

Step 2

Challenge your students to lay out headstones H_1, H_2, H_3 and H_4 somewhere in the school grounds using pegs. A sandy or grassy surface is good for doing this task. The square that the headstones make should have sides that are at least 3 metres long. Then mark where the treasure is with another peg.

(If this isn't possible, do it on paper in class, but doing this outside helps the students to get more intimately involved with the problem and will help them remember the event longer.)

What are the difficulties here? Let them plan their strategies in small groups before discussing their plans with the whole class.

Step 3

It shouldn't be hard to measure out the first side length, say 3 metres. This gives the positions H_1 and H_2 of the first two headstones. The question then is: how to find the position of the next two headstones?

The problem reduces to making a right angle at H_1 in order to locate H_4. Let the students discuss how this might be done. There are at least three ways. Give the students a chance to talk about their methods and then introduce whichever of the following they haven't thought of.

Method I: Use a T-square, protractor or some other device that has a right angle in it.

Method II: Use a 12-metre piece of string that has been tied off to form a loop. Mark in a point 3 metres from the join and another 4 metres from that. Then pull the looped string taut to form a right-angled triangle with sides 3, 4 and 5 metres, respectively. You should find a right angle where the 3-metre side meets the 4-metre side. (If your students already have some knowledge of Pythagoras' theorem, this could be a discussion point; however, it will not be introduced into the curriculum until Year 9 and could safely be left out of this exercise.)

Method III: Use rulers (straight lines) and compasses (these may be taut strings to draw circles) to construct a right angle at H_1. (See Step 4 to see how to do it and why it works.)

(There are other, more complex, methods that you can find by searching online.)

Give the students time to absorb these methods and try them out for themselves. What do they think is the easiest way to find the position of H_3? Which is the most difficult? Can they justify or prove each method?

Step 4

Encourage the students to discuss what they have done and the difficulties they found. What tricks did they pick up that would help if they had to do the task again? How 'square' was the shape they produced? (Errors will arise due to inaccuracies in their constructions as well as their use of string. How can these errors be minimised?)

Can they justify their method of making the square? We give the proofs below.

Method I: This should be clear. Put the T-square at H_1 so that it is perpendicular to H_1H_2 and measure 3 metres along the T-square. This will give the point H_4. Repeat this again at H_2 to give H_3 and then at H_3 to give H_4.

Method II: Since $3^2 + 4^2 = 5^2$ the 3, 4, 5 triangle is a right-angled triangle where the 5 side is the hypotenuse. This is not quite Pythagoras' theorem as it is usually stated, which is 'If a right-angled triangle has sides a, b and c, where c is the hypotenuse, then $a^2 + b^2 = c^2$'. But it is also true that if a triangle has sides a, b and c, where $a^2 + b^2 = c^2$, then the triangle is a right-angled triangle with hypotenuse c. The proof of this follows immediately from the Cosine Rule. Hence we can apply this second truth to see that the 3, 4, 5 triangle is a right-angled triangle with hypotenuse 5.

(Incidentally, Pythagoras' theorem can then be stated 'Let, a, b, c be the sides of a triangle. Then the triangle is a right-angled triangle *if and only if* $a^2 + b^2 = c^2$'. This is because we now know that the implication works both ways.)

So get the students to put the '3' side along H_1H_2 and pull the string taut to make the right-angled triangle H_2. The '4' side will then make a right angle with the '3' side. Going three metres along this '4' side will give them H_4.

Method III: We want to construct a right angle at the point H_1 on the line H_1H_2. First extend the line H_1H_2 to H_2', with H_1H_2' equal to H_1H_2. This equality can be assured by measuring the length H1H2 with the compass and then marking off the same length from H_1 on the side away from H2'. This is shown in Figure 2.1.

Figure 2.1: Constructing a right angle at H_1

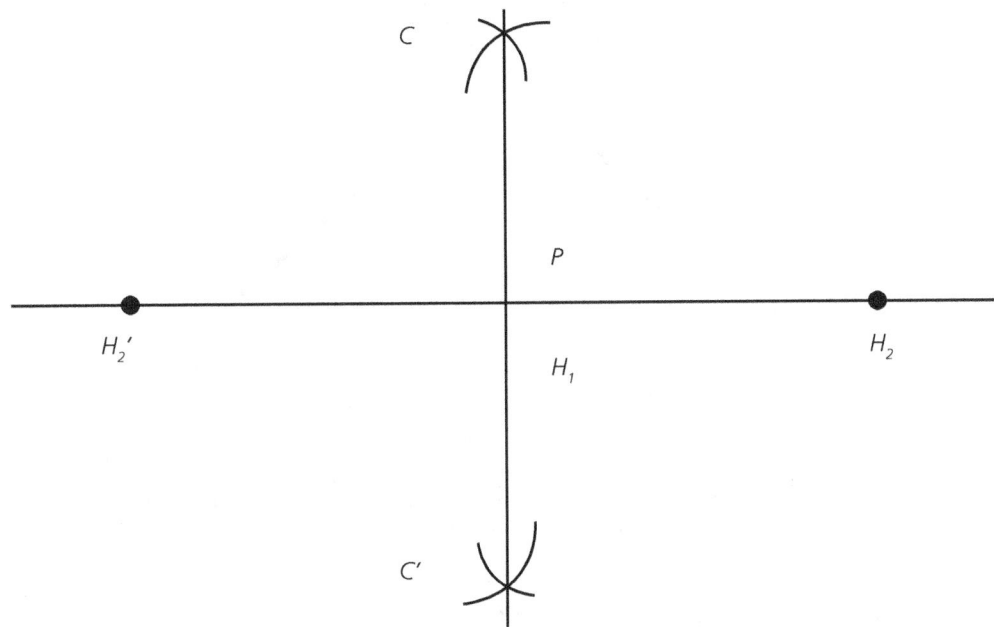

Now use the compass to draw two arcs of the same radius and centres H_2 and H_2', to meet to meet at some points C and C'. One of these points will be above $H_2'H_2$ and the other will be below. Call these points of intersection C and C', respectively. Join C to C' with a straight line. Suppose that P is the point where this line and $H_2'H_2$ intersect, then angle $H_2H_1C = H_2'H_1C' = 90°$ and $P = H_1$ is the midpoint of H_2H_2'.

Check this with a protractor. If a student's angle is a long way from being 90°, it might be useful for them to repeat the construction.

To justify this, first note that triangles H_1CH_2 and H_1CH_2' are congruent, SSS. ($H_1H_2 = H_1H_2'$ and $H_2C = H_2'C$ by construction; H_1C is in common.) So angle H_2H_1C = angle $H_2'H_1C$. But they are equal angles on a straight line, so they must both be right angles. Similarly, angle H_2H_1C' = angle $H_2'H_1C' = 90°$.

Hence CH_1C' is a straight line. But CPC' is a straight line joining C and C', so $P = H_1$ and so it is the midpoint of H_2H_2'.

Note: If in triangles ABC and DEF, AB = DE, AC = DF and BC = EF, then the two triangles are congruent. This type of congruence where the three sides of the triangles are compared is abbreviated to SSS.

Step 5

Now we justify that the diagonals intersect where the treasure is buried.

Figure 2.2: Why the method works

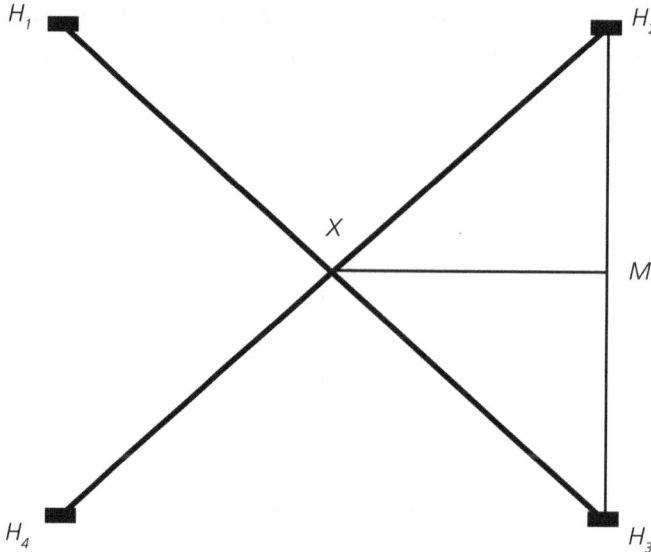

In the diagram, the headstones are H_1, H_2, H_3 and H_4; XM is parallel to H_1H_2; and X is where the strings cross.

Since XM is parallel to H_1H_2, angles XMH_2 and XMH_3 are right angles. Since angle XH_2M is the same as $H_4H_2H_3$, then angle XH_2M is 45°. Similarly, angle XH_3M is 45°. So XMH_2 is similar to triangle XMH_3. But XM is a common side to these two triangles, so the two triangles are congruent. Hence $XH_2 = XH_3$.

By a similar argument, $XH_2 = XH_4 = XH_1$.

Where to from here?

- What was harder, hiding the treasure or finding it? Why?
- Ask the students what construction methods they learnt here. Can they use these methods to make other shapes?
- What did they find was the easiest method to make a square? What was the hardest? Why?
- What other ways are there of hiding treasure? In these cases, is there a way to find where the treasure is?

Level 2: Three pegs

Problem

What if the treasure hiders decided to use three pegs instead of four? Now they plan to hide the treasure an equal distance from three pegs. Can it be done? How hard is it to find the treasure?

Problem steps

Step 1

Divide the students up into groups. Let them think about how they would hide the treasure and how they might find someone else's treasure.

In a whole class session, discuss what methods are practical. What methods do they think are the best and easiest, both for hiding the treasure and for finding it?

Step 2

Let every group hide their treasure so that no other group can see what they are doing. Then move the groups around to find the treasure of another group.

How successful were 'hiding' groups in placing their treasure equidistant from their pegs? How successful were the 'finding' groups? What methods did they use?

Step 3

There are at least two ways to hide the treasure:

1. **Method I (Hiding):** The simplest method is to take a piece of string (say 3 metres long), put one end at some point in the ground, pull it tight and draw a circle of radius 3 metres. Now choose any three points on the circle; these points are the vertices of a triangle. Bury the treasure at the centre of the circle, then delete all traces of the circle.

2. **Method II (Finding):** The first two pegs can go anywhere the group likes. Now take a fixed piece of string that is 3 metres long and make arcs of a circle of radius 3 and centres at the first peg (P_1) and second peg (P_2). Provided the distance P_1P_2 is less than 6 metres, the two arcs should meet at two points C_1 and C_2 (see Figure 2.3). (Note that there is no need to draw the whole circles; you just need arcs big enough to meet in two places.)

Figure 2.3: Placing the three points to hide the treasure

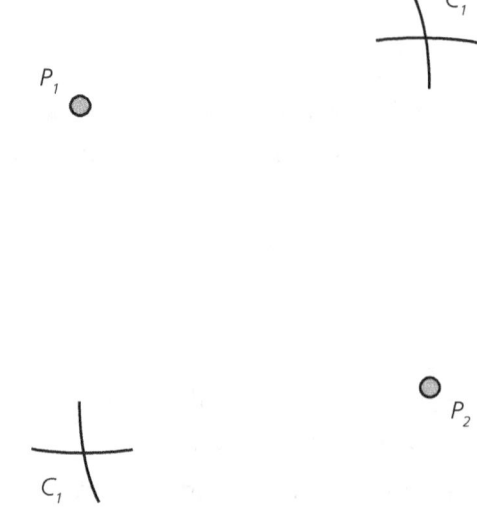

Now the groups might use their string to draw a circle of radius 3 metres with centre either C_1 or C_2. Any point on either of these circles, other than P_1 or P_2, can be used as the place to put a third peg. The centre (C_1 or C_2) is the place to put the treasure.

Step 4

There are at least two ways to find the treasure:

1. **Method A (Finding):** Take a string of any length and place it at P_1. Draw an arc that you think will meet with other arcs from P_2 and P_3 made with the same length of string. One of three things might happen.

   a. None of the arcs meet, in which case increase the length of the string.

   b. The three arcs meet at a single point. Bingo! You are lucky. You accidentally chose a string of length 3 metres and you hit the treasure in one go.

   c. The three arcs overlap. In this case, make the length of the string smaller and smaller until they only meet in one point.

2. **Method B (Finding):** Use the method of Figure 2.1 to produce a line that is perpendicular to a side and halfway along the side. Do that for two of the sides. These two perpendiculars intersect at the treasure.

Step 5

How accurate were the hiders? Was the treasure actually 3 metres from each peg?

How accurate were the finders? How many found the treasure the first time they used either Method A or Method B?

Step 6

It is proving time again.

For hiding: In Figure 2.4, the points C_1 and C_2 are 3 metres from P_1 and P_2 by construction. Any point on the circle drawn from these points is also 3 metres from them. So locate the treasure at C_1 (or C_2) and put the third peg on the circle drawn from C_1 (or C_2).

For finding: In Figure 2.4, suppose that G is any point on the perpendicular bisector of P_1P_2. Then triangles P_1HG and P_2HG are congruent (SAS). So $P_1G = P_2G$.

Figure 2.4: Finding the treasure

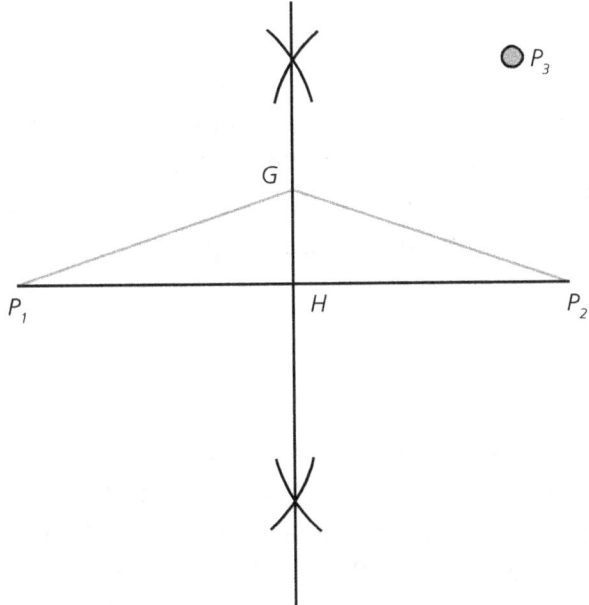

If we now construct the perpendicular bisector of P_2P_3, this will meet the line HG, at some point J. By the argument above, $P_1J = P_2J = P_3J$. The treasure must be at J.

Where to from here?

- Note that the circle that the pegs are on is called the *circumcircle* of the triangle that has three peg points as vertices. The centre of that circle, where the treasure is buried, is called the *circumcentre*.
- Does the treasure *always* lie inside the triangle made by the pegs? If your students think it does, ask them to prove it; if they think it doesn't, they need to show an example of a triangle where the treasure is not inside it. It turns out that for acute-angled triangles the treasure will lie inside the triangle; for a right-angled triangle the treasure is at the centre of the hypotenuse; and for an obtuse-angled triangle the treasure will be outside the triangle. This can be seen by choosing appropriate points for the third peg in Figure 2.4.
- How will your students hide the triangle if they just took any three points at random?

Level 3: Any four points

Problem

Any three points to choose will lead to a unique point for the treasure. Four points that are corners of a square will also lead to a unique point for the treasure, where each peg is the same distance away from the treasure. Is it true that *any* four points we choose for the pegs (not necessarily on a square) will lead to a unique point for the treasure?

Problem steps

Step 1

Start this activity by getting the whole class together. What do your students think about the problem? Let them think about it for a while, then ask them to vote 'yes' if they think that any four points will give a unique treasure point, or 'no' if there are some sets of four points that won't give a treasure point. Record the overall results.

Remind them that if they voted 'yes', they have to justify that it works for all sets of four points. If they voted 'no', they only have to produce one example to show that it doesn't work. Then allow a class discussion.

Step 2

Why not start off with some quadrilateral (apart from a square) whose vertices your students know are equidistant from a given point? What suggestions does your class have?

Someone will no doubt suggest a rectangle. Can they prove their suggestion? The argument is almost the same as that in Level 1, Step 5 after Figure 2.2 (page 64). You just need to appeal to properties of the rectangle to get angles XH_2M and XH_3M to be equal.

Can they think of any other quadrilaterals that can be used to bury treasure? Let the class discuss any further suggestions and prove that they are right (or wrong).

Step 3

On the other hand, are there quadrilaterals that *definitely* won't work? Such sets of four points exist. Can they think of any?

One extreme example is the 'flat' rhombus in Figure 2.5.

Figure 2.5: Some quadrilaterals of pegs cannot be used to bury treasure

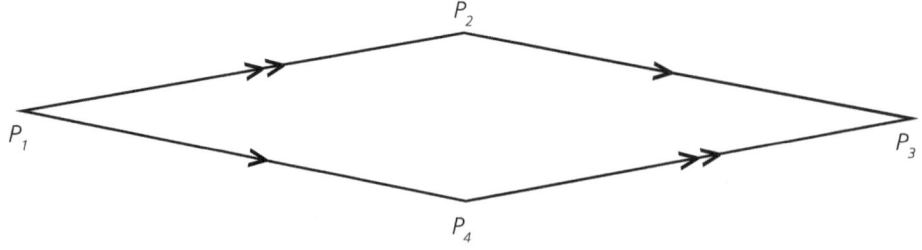

Is it true that the pegs in the rhombus shape of Figure 2.5 aren't useful for hiding treasure? Can your students show this in more than one way? The students should work on this in their groups.

There are a couple of ways to do this. First, the three perpendicular bisectors of P_1P_2, P_2P_3 and P_3P_4 don't meet in a point at all! Second, they might notice that the only points that are the same distance from P_1 and P_3 must lie on the line through P_2P_4. So the treasure point has to be halfway between P_2 and P_4. But this is clearly less than the distance from P_1 to any point on P_2P_4. So this quadrilateral isn't a good one to use in hiding treasure.

So is there any rhombus at all (other than a square) that has a treasure point? At the moment it is hard to see this unless you have someone who can animate the situation. This would require the meeting points of the pairs of the three bisectors we have talked about changing as the shape of the rhombus changes.

Step 4

Now we take another approach. By now many of your students may have realised that if we take a circle and choose any four points on its circumference, then we have a good treasure-inducing quadrilateral. Clearly the opposite is true; if we have four points that are equidistant from a given point they must lie on a circle. So we can focus on quadrilaterals that live on circles. We call these *cyclic quadrilaterals*. Do such quadrilaterals have a simple property that characterises them as cyclic? Or do we have to find where the centre is to be sure that they are cyclic? What does your class think?

Basically, your 'raw' quadrilateral only has edges and angles. Get your students to draw a few quadrilaterals and measure the sides and the angles. Make a table to see if anything consistent is happening. Get them to test any conjectures with new examples. If any conjecture seems to hold, get them to try to prove it.

Step 5

The key here is that the opposite angles in a cyclic quadrilateral add to 180°. So is it true that a quadrilateral is cyclic *if and only if* its opposite internal angles add up to 180°?

This result requires the students to know that the angle at the centre of a circle is twice the angle on the circumference that subtends the same angle.

Look at Figure 2.6. Let O be the centre of the circle. Then the internal angle at P_2 is a half of the angle at O on the P_4 side. Similarly, the internal angle at P_4 is a half of the angle at O on the P_2 side. But the sum of the angles at O is 360°. So the sum of the internal angles at P_2 and P_4 is half of this: i.e. 180°.

Figure 2.6: The internal angles at P_2 and P_4 add up to 180°

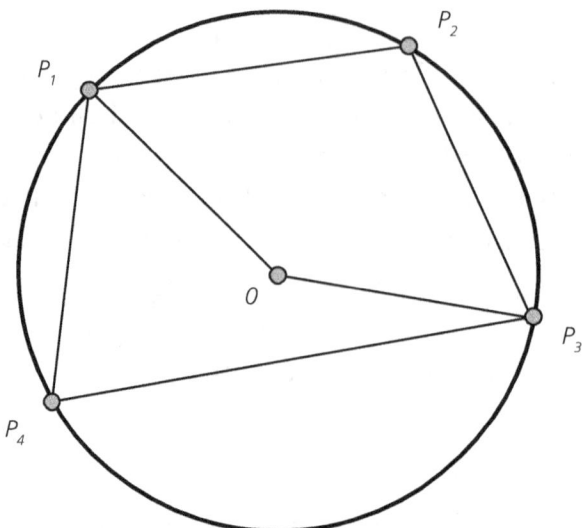

We have now shown that a cyclic quadrilateral has opposite internal angles that sum to 180°. But does it work the other way? Is it true that if the opposite internal angles of a quadrilateral add up to 180° then it is cyclic? (We outline this proof in the 'Where to from here?' section.)

Step 6

Which rhombi are cyclic? Let your class experiment again with this one.

Because of the symmetry of a rhombus, its internal and opposite angles are equal. Suppose that these angles are $x°$. For the rhombus to be cyclic, $x + x$ must be 180. Therefore $x = 90$ and the rhombus must be a square. So most (proper) rhombi are not cyclic quadrilaterals.

Where to from here?

- We can outline the proof that if the opposite internal angles of a quadrilateral add up to 180° then it is cyclic. Suppose that these angles are a, b, $180° - a$ and $180° - b$. Take three of the vertices of this quadrilateral so that they contain the angle a. We then know those three vertices have a unique circle through them. Suppose that this circle does not go through the fourth vertex of the original quadrilateral. Now open up two of the angles of this triangle other than a to make the angles b and $180° - b$. These form a fourth angle on the circle that has to have an angle of $180° - a$. Show that this angle must be at the vertex where we had the original angle of $180° - a$.

- Can your students show that a concave quadrilateral isn't cyclic?

- A cyclic polygon is any polygon whose vertices lie on a circle. What figures do your students know that are cyclic? Do they have any interesting internal angle properties?

Level 4: Cyclic hexagons

Problem

Suppose that you wanted to bury treasure using six pegs. How would you know where to put them?

Problem steps

Step 1

This isn't too hard if there is no restriction on where to put the pegs—just find a circle and put six pegs on its boundary. Of course, this method applies to any sized polygon.

But what would your class do if you wanted to use six pegs in the shape of a *regular* hexagon?

Step 2

To do this, draw a circle with any radius. Remind the students that from here, it is important to keep the compass fixed at that radius.

Place the point of the compass anywhere on the circle that has been drawn. Call this point P. Draw out enough of the circle centred at P, so that you can see the two points Q and U where this second circle hits the original one. Now draw enough of a third circle, centre Q, so that the two points where the third circle cuts the first one can be seen. One of these points is P. Call the other R.

Repeat this construction with a circle at centre R to produce the intersection points S and Q; repeat with a circle centre S to give the new intersection point T. Now use a straight edge, ruler or piece of wood or whatever, to join the six points P, Q, R, S, T and U as shown in Figure 2.7.

Figure 2.7: Constructing a regular hexagon

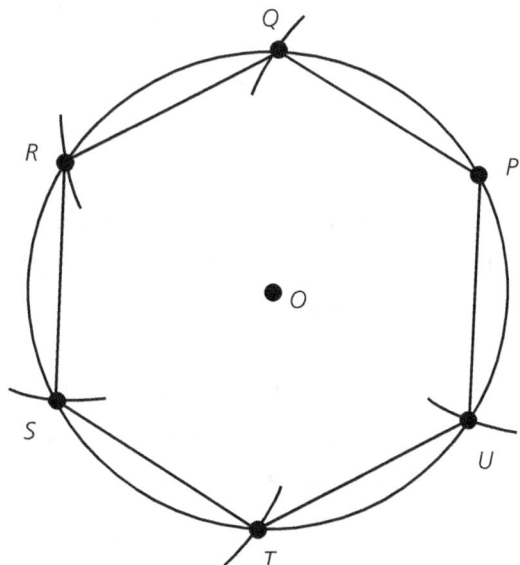

Can you prove that the hexagon constructed above is indeed a regular hexagon?

Step 3

Let O be the centre of the first circle. Then $OP = PQ = QO$, because they are all equal to the radius of the first circle. Similarly, $QR = RS = ST = TU = UP$. Since all of the sides are equal, $PQRSTU$ is a regular hexagon.

But how do you find the treasure in the case of pegs at the vertices of a regular hexagon?

Step 4

Are there any methods that have been used on other shapes that will help with the regular hexagon?

Your students should experiment to see what they can come up with. It turns out that the method of Level 1 involving the intersection point of the diagonals of the hexagon, and the method of Level 2, which involves the perpendicular bisectors of sides, will both work here.

Can your students verify these?

Step 5

Intersection points: Thinking about the drawing in Figure 2.8, we know that $PQ = QR$. So PQ and QR subtend the same angles both on the circumference of the circle and at the centre.

Figure 2.8: 60° angles

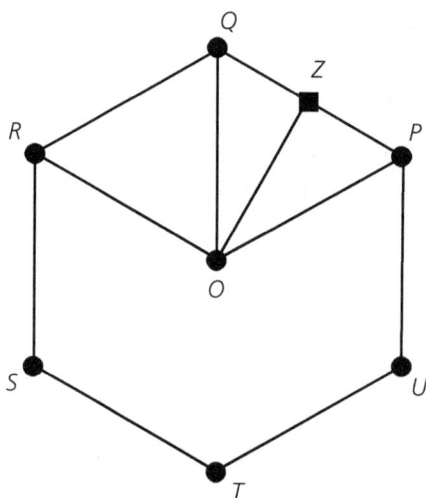

In that case, all six angles POQ, QOR, ROS, SOT, TOU and UOP are equal. Since these six angles add up to 360°, they must all equal 60°. So angle POS is 60° + 60° + 60°. Hence $PO + OS$ is the straight line PS. This means that all diagonals go through O.

Perpendicular bisectors: In Figure 2.8, consider the line from O to PQ such that $PZ = ZQ$. Since triangles POZ, QOZ are congruent (SSS), the angles at Z are equal. Hence angle PZO = 90°. So the perpendicular bisectors of the sides of the regular hexagon meet at the centre O.

Did your students find any other methods? Can they justify them?

Step 6

From what we saw in Step 5, the centre of the circumcircle of the regular hexagon is also the centre of a circle for which the sides are tangents. (These tangents don't need to touch the sides at their midpoints, though they do for regular hexagons.) Such circles are called *incircles*. Ask your students to construct a regular hexagon and its incircle using ruler and compass.

What is the ratio of the circumradius to the *inradius* of a regular hexagon?

Step 7

Figure 2.9 shows the circumradius r and the inradius r' of a regular hexagon.

Figure 2.9: Circumradius and inradius of a regular hexagon

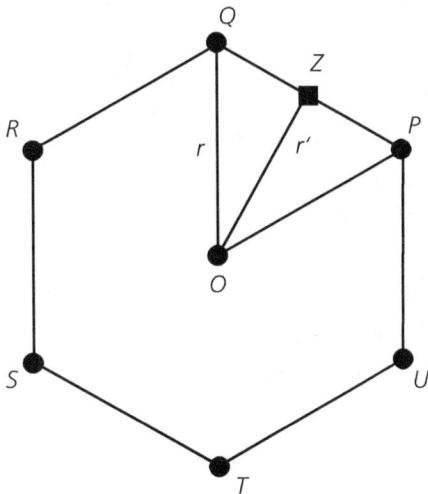

In triangle POZ, $PZ = \frac{r}{2}$. Since the triangle is a right-angled triangle, $r' = (\frac{\sqrt{3}r}{2})$.

Where to from here?

- No matter what, it is very easy to find the treasure given six pegs that are at the vertices of a regular hexagon. So this doesn't seem a very good shape to leave your pegs in. What is the most difficult shape for someone to steal the treasure from?
- An extension activity, 'Incircles of triangles', can be downloaded from the series website.

CHAPTER 5: TESSELLATIONS

Initial problem

Kylie is designing the floor covering for a new exhibition centre. She decides to cover the floor of the main hall with polygonal tiles that are all the same. What polygons could she use to do this?

Background information

People have always decorated their dwellings, and mosaics are well known from classical times. Roman mosaics were often pictures made of small pieces of pottery of different colours. In more modern times we usually cover floors and walls with tiles made from one shape, which may be a square or some other regular shape.

This chapter is an extension of the 'Horrible Hal's humungous hall' activity in *Creative Activities in Mathematics Book 1*. In the current activity we look at shapes that will tile the plane, and we do this through exploration. The first two levels are given over to tiling by polygons. In Level 1 the polygons are regular; Year 7 and 8 students should be able to tackle this level.

In Level 2 we look at what other polygons might tessellate. Many Year 7 students as well as all Year 8 students should be able to tackle Level 2.

The second two levels involve *polyominoes*—shapes that are made by joining together squares along their edges. For Level 3 the emphasis is on listing polyominoes and seeing which of these tessellate; this level is ideal for Year 9 students but is open to experimentation by Year 8 students.

In Level 4 we look at some polyominoes that tile rectangles. Again, this is best for Year 9, but many Year 8 students should be able to make some progress.

Table 2.3: Australian Curriculum content descriptions for the *Tessellations* activity

Activity level	Problem	Content descriptions
1	Regular polygons	*Year 7* Demonstrate that the angle sum of a triangle is 180° and use this to find the angle sum of a quadrilateral (ACMMG166)
2	Polygons	*Year 7* Classify triangles according to their side and angle properties and describe quadrilaterals (ACMMG165) Describe translations, reflections in an axis, and rotations of multiples of 90° on the Cartesian plane using coordinates. Identify line and rotational symmetries (ACMMG181) *Year 8* Establish properties of quadrilaterals using congruent triangles and angle properties, and solve related numerical problems using reasoning (ACMMG202)
3	Polyominoes	*Year 9* Apply logical reasoning, including the use of congruence and similarity, to proofs and numerical exercises involving plane shapes (ACMMG244)
4	Tiling with polyominoes	*Year 9* ACMMG244 (see above)

Problem aims

» Increasing students' geometric intuition, especially of polygons
» Making and proving conjectures

Key concepts:

» Polygons and regular polygons
» Interior angles in a polygon
» Angles associated with transversals of parallel lines
» Polyminoes

Possible heuristics/strategies

» Trial and error

Concrete materials

» Paper and scissors

Special notes

Tiling (tessellation): Tiling is a way of covering the plane with shapes (tiles or *tessella*) so that there are no gaps between the shapes. Often the tiling is done with one shape.

Polyomino: A polyomino is a shape created by joining a finite number of squares without overlap so that edges fit exactly to edges).

Level 1: Regular polygons

Problem

Kylie is designing the floor covering for a new exhibition centre. She decides to cover the floor of the main hall with polygonal tiles that are all the same. What polygons could she use to do this?

Problem steps

Before tackling the problem, we start with steps that explore the concepts and give students a foundation for their work.

Note that this level has some proofs in that you may want to avoid for your class. The substance of the tiling can be done without these proofs, though it would be good if you could give the students some idea of why things are true.

Step 1

Start with the basics: What is a polygon? What is a regular polygon? Can your students give examples of each?

Get the class to discuss the two questions. What are 3-, 4-, 5- and 6-sided polygons called? What about 3- and 4-sided regular polygons?

Come up with definitions of polygons and regular polygons and test these definitions against the examples you have discussed. Get the students to check the definitions using an online mathematical dictionary.

Draw or collect pictures of the small regular polygons with less than or equal to 10 sides. What are the names of these polygons? Display them on the classroom wall.

Step 2

Next, ask the class for examples of real-life situations where surfaces are tiled with regular polygons, and what effect the boundaries of the surfaces have. Explain that, for this discussion, *tiling* means a situation where the tiles fill the space completely and to the edge of the space.

For examples, in a shower tiled with squares, the walls are covered in square tiles with no gaps between them and with complete tiles round the edges of the wall. On the other hand, in a wall made with rectangular bricks, there are always part-bricks at the edges—so the bricks do *not* tile the wall.

Draw or collect pictures of rectangles that have been tiled with polygonal shapes.

Step 3

Imagine now that we wanted to tile a very big floor—in fact, a floor that goes off to infinity in all directions. Can we do this with tiles that are all in the shape of the same *regular* polygon?

Do this systematically by looking at equilateral triangles first and then increasing the number of sides the regular polygon has. Give your students several equilateral triangles (or get them to construct them and cut them out) that are the same size and let them experiment to show that the plane can be tiled with these regular polygons. Obviously they can't show the whole tiling of an infinite plane, but they can show that it can be done by looking at a configuration such as the one in Figure 2.10.

Figure 2.10: Tiling with equilateral triangles

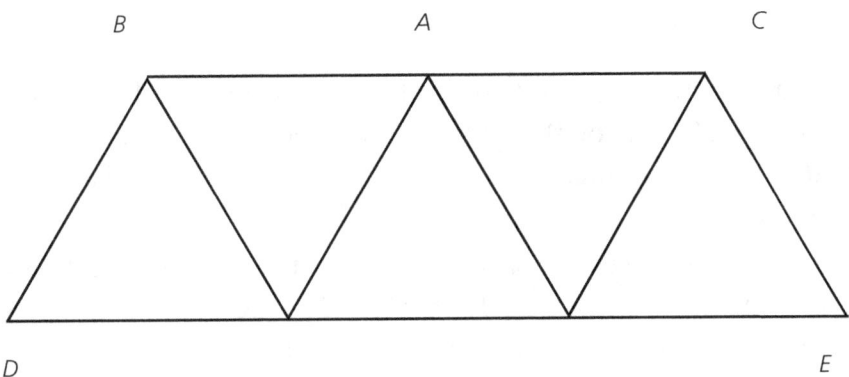

At a point such as A, three equilateral triangles come together. As the angles of an equilateral triangle are 60°, then BAC is a straight line. The same can be said for DE.

So we can fit a line of equilateral triangles together that has parallel sides so that there are no gaps between these sides. Then we can put two of these lines of triangles together with no gaps between them. (The strip approach of Figure 2.10 will be reused throughout the activity to show how many shapes will tile the plane.)

(Note that we justify the sizes of the interior angles of a regular polygon with n sides in Step 5. You may want to do this before you do the previous steps. Alternatively you may just want to assume what the sizes of the interior angles are. This will depend on your class.)

Step 4

Continue efforts to tile the plane with regular polygons by experimenting with squares, pentagons and hexagons. Your students should be able to tile the plane with squares using the strip method of Figure 2.10.

There is a problem with regular pentagons; they don't fit together without leaving gaps. This is because the interior angles of a regular pentagon are 108° and 108° doesn't go exactly into 360°. This is crucial in two ways. First, the whole point of a tiling or tessellation is that all of the points on the floor or plane can be covered by the tiles, with no gaps. Second, we'll use this test of angles adding up to 360° again as the activity progresses.

While they can't tile with regular pentagons, the students should be able to make the regular hexagon tile the plane. Encourage them to explain why their tilings work. We give an argument below that they might use. This is based on Figure 2.11.

Figure 2.11: Strips of regular hexagons tile the plane

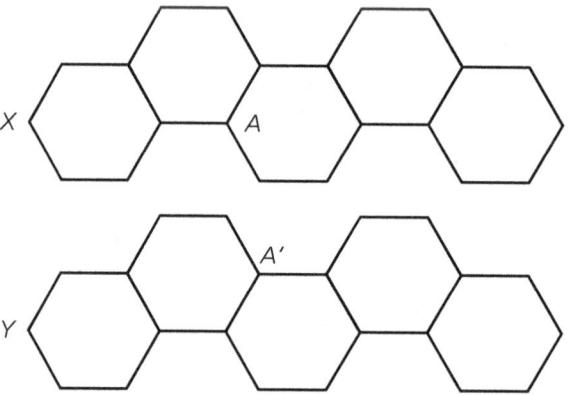

In the diagram, the strip X goes off forever to the right and left. The angle on the inside of A is $120° + 120° = 240°$ and on the outside of A is $360° - 240° = 120°$. The angles at any corner along the strip X have angles of 120° and 240°. Sometimes the larger angle is inside the strip; sometimes it is outside.

The infinite strip Y is exactly the same as X. These two strips fit together to fill up the gaps left by X because the outside angle at A' is 120°. Continuing in this way with an infinite number of strips like X, we can tile the whole plane.

Step 5

We know that we can tile the plane with regular n-sided polygons for $n = 3, 4$ and 6. But can we do the same with any other value of n?

We have seen that the problem with $n = 5$ is that the interior angles are not divisors of 360° so there have to be gaps. So we can deduce that, if we know the interior angle of an n-sided polygon, we can tell whether it can be tiled or not.

To determine that angle, we first have to ask: What is the sum of the interior angles of a regular polygon with n sides?

We can start by first proving that the sum of the interior angles of *any* triangle is 180°. This can be introduced by everyone in the class cutting out a triangle from paper and then tearing off the corners. The three torn-off corner angles can be put together to look like a straight line. Alternatively, the class can draw a triangle and measure its three angles so that they can see that 180° *might* be the right answer.

However, these are just demonstrations that show it is feasible that the sum is 180°. There is no way that your class can produce and measure *all* possible triangles to make sure that the result is true for *every* triangle. On the other hand, if they know about equal angles associated with parallel lines, they can be shown the proof below.

In Figure 2.12 we take any triangle *ABC* with interior angles *a*, *b* and *c*. Now extend the base line *AB* and add a line *CD* that is parallel to *AB*.

Figure 2.12: Proof that the sum of the interior angles of a triangle is 180°

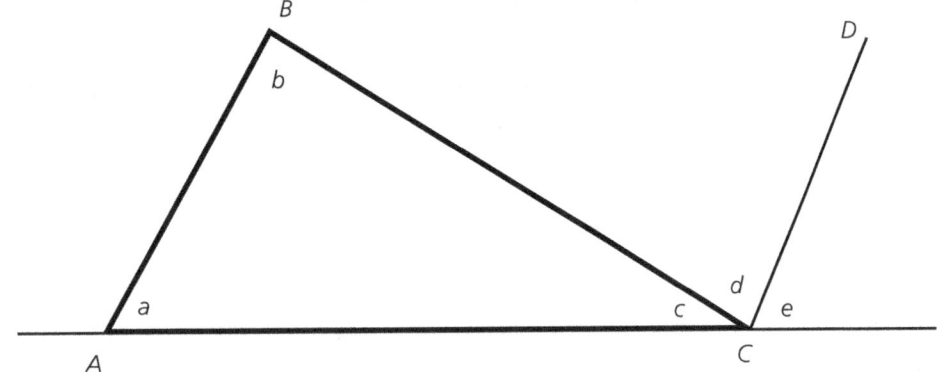

We see that $a = e$ because they are corresponding angles on parallel lines and $b = d$ because they are alternate angles on parallel lines. This means that $c + d + e = 180°$ because they make up the line AC extended. Therefore $a + b + c = e + d + c = 180°$.

Going back to equilateral triangles, the sum of the interior angles has to be 180°. As a result, because we know that each angle is equal, we know that they are equal to $180 \div 3 = 60°$.

From here it is just a matter of dividing the regular polygon up into triangles, but let your students think about this first and come up with the idea themselves. For the square, this is done simply by drawing in a diagonal. Figure 2.13 shows how it is done for a regular hexagon, and by extension for a polygon with a larger number of sides.

Figure 2.13: Dividing a regular hexagon into triangles

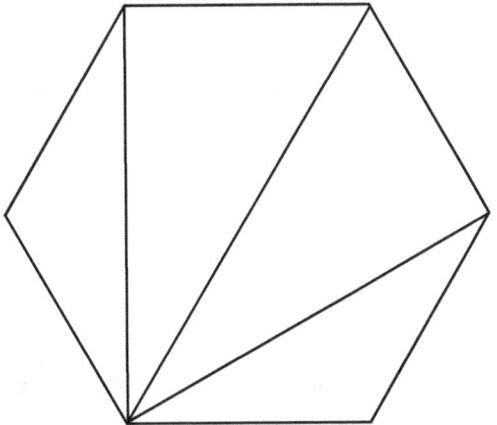

If we add up all of the angles of the triangles, we get all of the interior angles of the hexagon, which thus add up to $4 \times 180° = 720°$.

Let the students now find the sum of the interior angles of an n-sided regular polygon with $n = 7, 8$ and 9.

So for any number of sides, the number of triangles at the vertex of a regular n-gon is two less than the number of sides or $n - 2$. This means that the sum of the interior angles is $180°(n - 2)$.

What is the size of *any* interior angle in such a polygon? It is just $180°(n - 2)$ divided by n.

Step 6

Why are regular 3-, 4- and 6-sided polygons the only ones that will tile the plane?

Remember that we have to fill each corner. So if we try to fit tiles so that all of the vertices of the tiles fit together at a point, then we need $180(n - 2)$ n to divide 360. As a result, $360 \div [\frac{180(n-2)}{n}]$ has to be a whole number. But:

$$360 \div \frac{180(n-2)}{n} = \frac{2n}{n-2}$$

$$= \frac{2=4}{n-2}$$

(This extra equality is a trick that makes it easier to see that you only need to try a small number of values of n.)

If this is to be a whole number, then $n - 2$ *must* be a factor of 4. So $n - 2$ can only be 1, 2 or 4. (Get your class to try other possibilities until they are satisfied.) This gives $n = 3, 4$ or 6.

If we don't fit tiles together at a vertex, the only other way that we could fit them together is shown in Figure 2.14 below, where parts of the sides of two polygons fit together. But then we have to have a gap at A, because for $n > 4$, the interior angles of a regular polygon are bigger than 90°.

Figure 2.14: Another way to fit tiles together

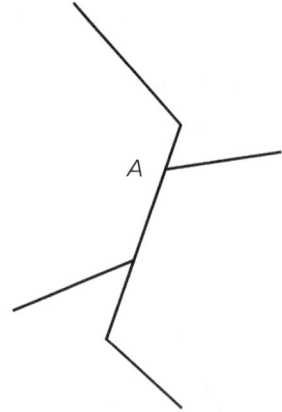

Where to from here?

- Can we use two different regular polygonal tiles to tile the plane?
- How about more than two regular polygonal tiles?

Level 2: Polygons

Problem

What other kinds of polygonal tiles can Kylie use to tile the main hall?

(Here we drop the condition that the tiles have to be regular.)

Problem steps

This is a lot to try to do all at once. We suggest that your students use a guillotine to cut up polygons of the same size and then try to fit them together to tile the plane. This can also be done using a computer—draw any triangle, copy it several times and then fit the triangles together.

Using trial and error to find a solution is rather hit or miss, so after a while it may appear more useful to be systematic. So start small and cut up several copies of the same triangle. Can they tile the plane with these? Then progress to polygons with four sides and so on.

Step 1

What triangles will tile the plane?

Here is an example of an acute and an obtuse triangular tiling (Figure 2.15). It is based on the equilateral tiling from Level 1. In both cases we make a strip of triangles that fit together between a pair of parallel lines.

Figure 2.15: Strips of triangles that lead to a tiling of the plane

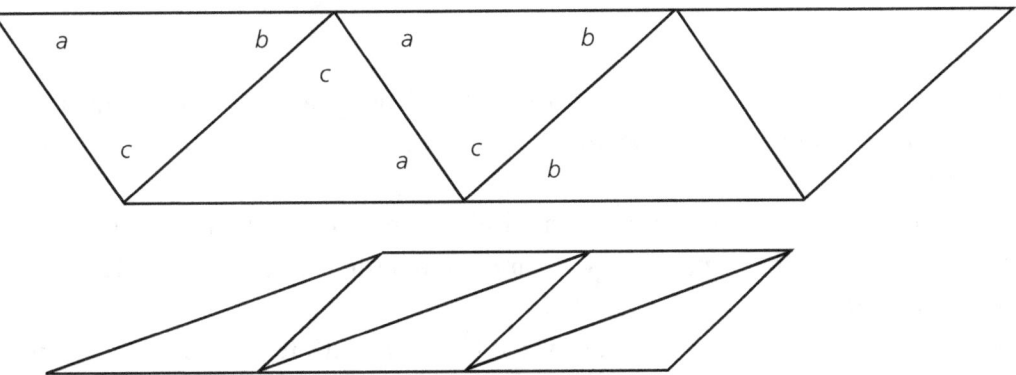

Step 2

Ask the class: How we can be sure that there are no gaps in the tilings in Step 1?

Get the students to use protractors to measure the angles of the triangles and then demonstrate that they line up as in the first drawing in Figure 2.15. (The second drawing works the same way.)

We have shown these angles as a, b and c. Because $a + b + c = 180°$, the sum of the interior angles of a triangle (see Level 1), we know that the strips have edges that make a continual infinite line.

We now know the following:

Theorem: Any triangular-shaped tile will tile the plane.

Step 3

Is the same result true for any four-sided polygon (quadrilateral)?

It is again time for experimentation with whatever quadrilateral shapes your students can imagine. This can be done using physical shapes that they cut out or virtual ones on a computer. This experimentation is an important part of the exercise; it gives students the opportunity to develop their intuition and see things for themselves, both of which aid understanding and hence learning.

Now get the class to produce all of the types of quadrilaterals they can imagine. Figure 2.16 shows what has to be considered. Shapes that we haven't shown here can be obtained from those that have. For instance, the square is a special case of a rectangle, the rhombus is a special parallelogram and a kite is a special case of the fourth quadrilateral.

Figure 2.16: The basic types of quadrilaterals

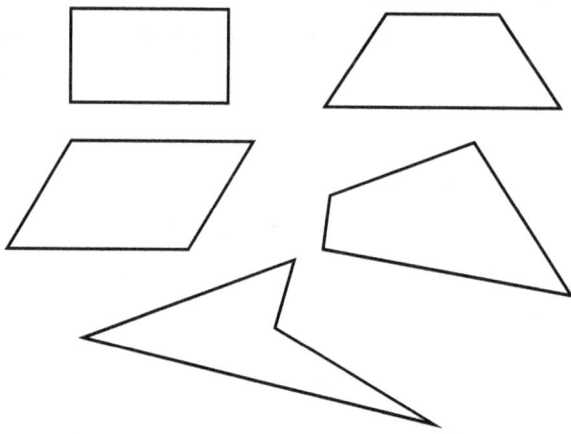

Step 4

Get the class to work in groups to see how many of the quadrilaterals in Figure 2.16 they can use to tile the plane. When they have done as much as they can, get them together to discuss their results.

Bricks won't tile a finite wall unless the dimensions are just right, but they *will* tile an infinite wall. Just take the tiling of a square and stretch the whole plane until all the squares are the same rectangle.

The trapezium tiles can be proven to work for tiling with the method shown in Figure 2.17. Again we take a strip approach. Alternate trapezia in the strip are obtained from each other by rotations.

Figure 2.17: A strip on which to base a tiling by trapezia

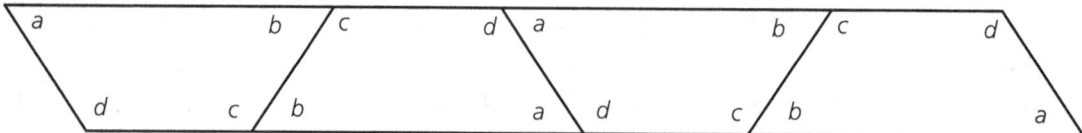

Because opposite sides of a trapezium are parallel, angles *a* and *d* and *b* and *c* are complementary. This forces the two opposite sides of this strip of trapezia to be straight lines. Putting two strips together will therefore leave no gaps between the strips. We can then tile the plane using an infinite number of strips.

Parallelograms will tile the plane in exactly the same way.

Things get slightly more difficult, and therefore more interesting, in the fourth case: the kite-like shape. Many students will believe that this quadrilateral on our list won't tile the plane, but encourage them to give it a try. They should try to keep the strip idea, but realise that the strips here won't have parallel edges; instead, they are more like the strips we used in the case of the regular hexagon in Level 1. We show these strips and how they fit together in Figure 2.18.

Figure 2.18: Tiling with an awkward quadrilateral

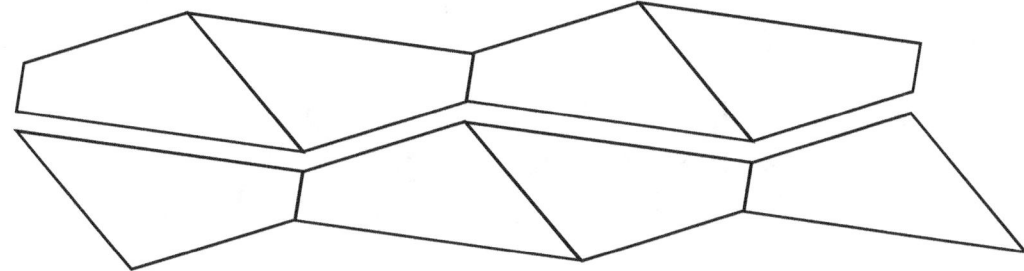

The crucial thing here is that the interior angles of a quadrilateral add up to 360°. Your students should make sure that all of the angles of this quadrilateral meet at the various points where the strips meet. Some specific examples with given angles might help them to see the general case.

A similar thing happens in the last case for the concave quadrilateral (see Figure 2.19), but it may be harder to discover because of the interior angle that is bigger than 180°. We show the crucial strip below and leave your students to put other strips with it to tile the plane. They should also be able by now to show that all of the tiles do fit together.

Figure 2.19: A concave quadrilateral stripped for action

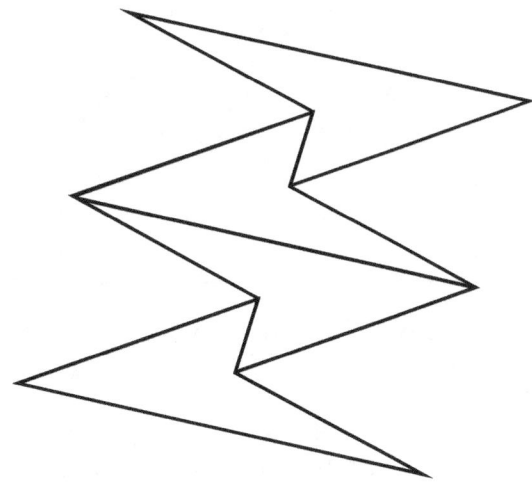

As the result of this work, your students have proven that all quadrilaterals will do for a tessellation. Kylie might take note.

Step 5

So what about pentagons? We had no luck with the regular pentagon; does that suggest that no pentagon can be used to tile the plane? What do your students think?

This is the cue for your students to experiment. It turns out that there are many pentagons that will tile the plane. Here is one example: the 'side-of-a-house' pentagon (Figure 2.20). In fact, there are 15 distinct pentagons that tile the plane – and the 15th was only discovered in mid-2015! (See the series website for news on this discovery.)

Figure 2.20: A 'side-of-a-house' pentagon tile that will tile

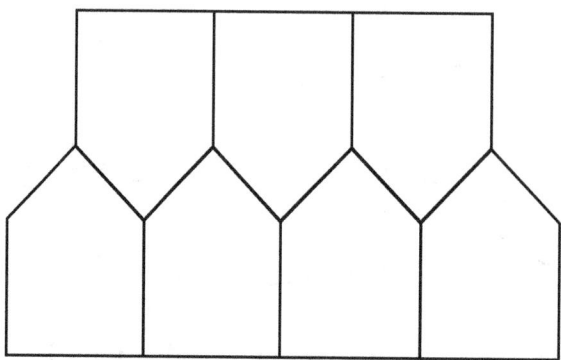

We leave the justifying to your students, as the ideas they need are ones they have already used.

Step 6

So is there a tile for some n-sided polygon for every $n > 2$?

Once more the students should experiment. There are many solutions for each value of n. We show one that will work for an odd value of n in Figure 2.21. This generalises the 'side-of-a-house' tile. You just need to incorporate an even number of 'up' and 'down' pieces for the roof.

Figure 2.21: An odd tile

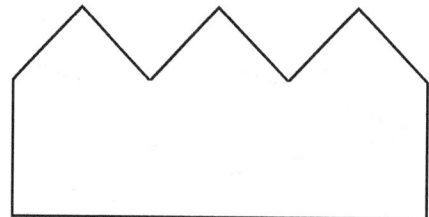

You can find tiles for even values of n by having one more 'up' piece than 'down'.

Where to from here?

- Why not bring some symmetry into this? Can your students find tiles with n sides that have rotational symmetry and tile the plane? What symmetry can they find in the tessellations that we have in this level?
- Is it possible to find a pentagonal tiling where the pentagon has no parallel sides?
- What do your students think is the 'nicest' tile for tessellating that has been used so far?
- Can the class make some Escher-type tessellations based on what has been done in this level?

Level 3: Polyominoes

Problem

A *polyomino* is a shape made using *n* squares, such that all squares are joined to neighbouring squares along the whole of the two joining edges.

Apart from the square itself, the simplest polyomino has two squares and is called a *domino* (Figure 2.22).

Figure 2.22: A domino—a polyomino with two squares

There are two *triominoes*—polyominoes with three squares—shown in Figure 2.23.

Figure 2.23: The two triominoes

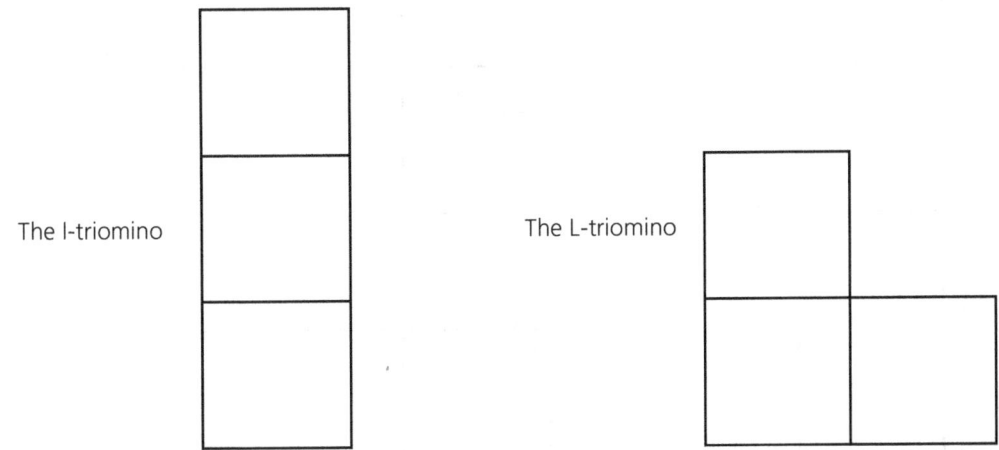

Can the students find all of the *tetrominoes* (polyominoes with four squares) and *pentominoes* (with five squares)?

Problem steps

Step 1

This will require your students to use systematic trial and error. It also requires a class decision: are the two tetrominoes in Figure 2.24 the same?

Figure 2.24: Two tetrominoes we will call the same

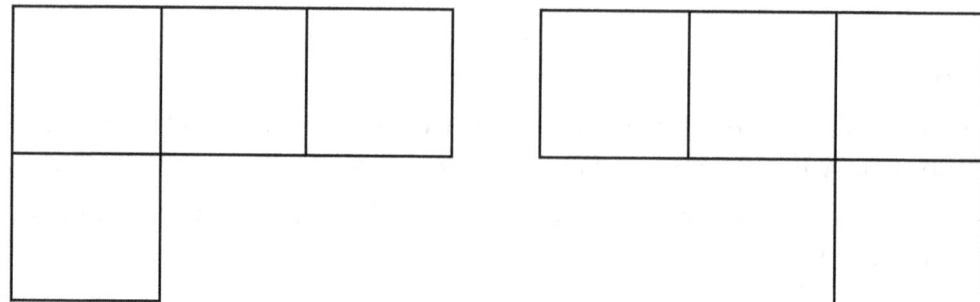

We assume that they are the same as it means we have fewer cases to consider. In general, any two polyominoes that can be obtained from each other by a translation, rotation, reflection (like the two above) or glide reflection will be considered to be the same. (These are sometimes called free polyominoes.) This means that there are five tetrominoes, as shown in Figure 2.25.

Figure 2.25: The five tetrominoes

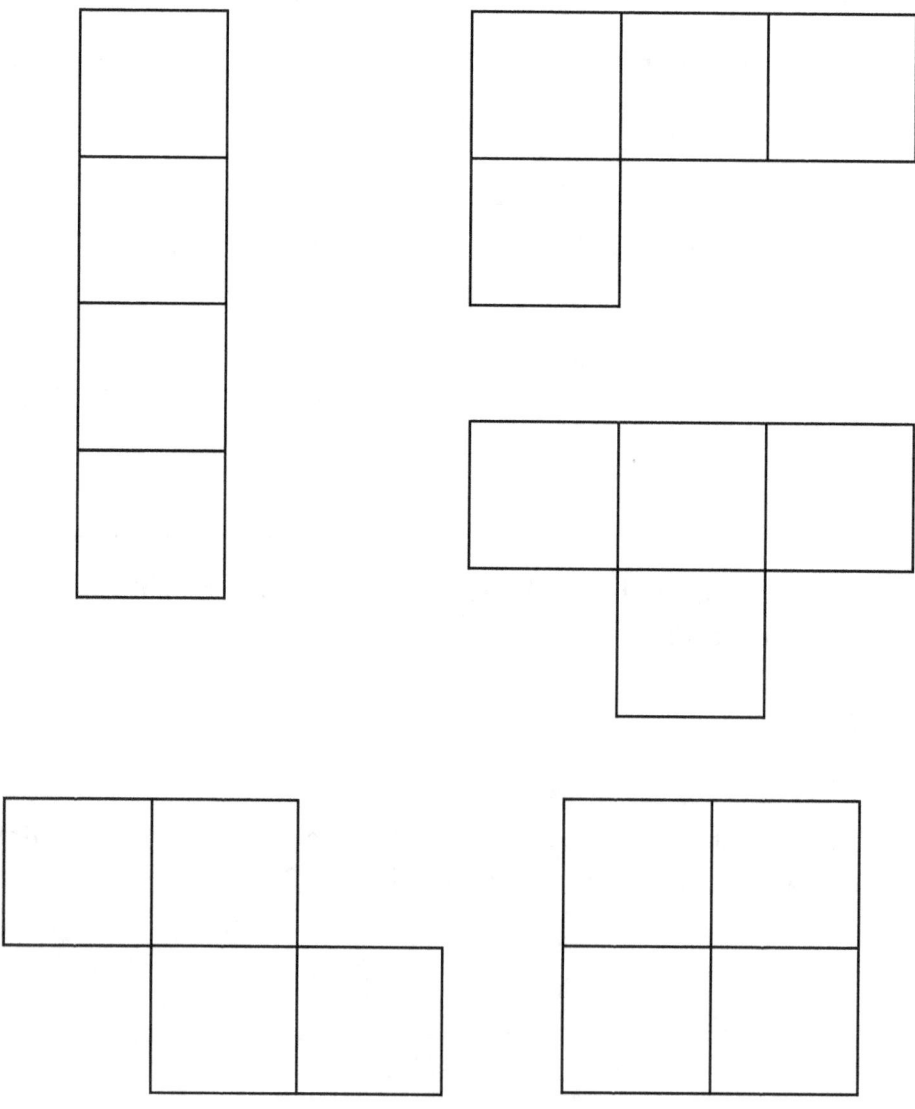

There are 12 free pentominoes—too many to reproduce here, but you can find them online via the series website.

Step 2

Which tetrominoes can tile the plane?

They all can, using the strip approach that students should now be familiar with. Figure 2.26 shows the only one that might worry or confuse your students.

Figure 2.26: A strip tiling of an awkward tetromino

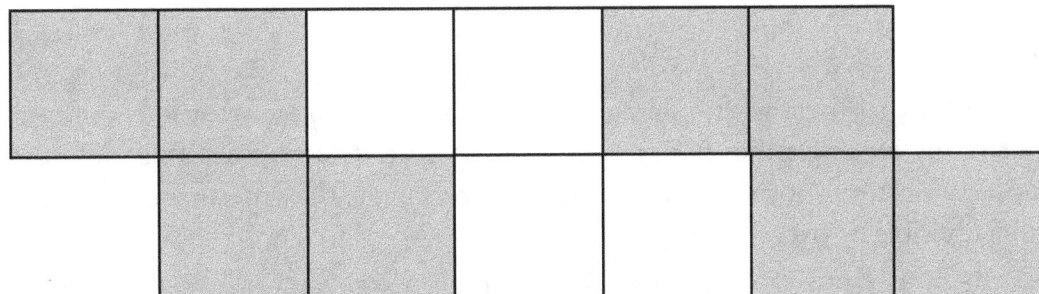

It doesn't matter that one square 'hangs off' each end. We can eventually cover all the parts of a strip two squares wide by adding enough of these tetrominoes at either end. This is where the infinite has an advantage over the finite.

Step 3

Now get your students to investigate which pentominoes will tile the plane.

Once again ... they all do. It might be a challenge for the students to work them all out, though they could do this over a period of time.

Where to from here?

- What is the smallest polyomino that will *not* tile the plane? Well, it isn't a hexomino—they all tile. It actually turns out that it is the seven-square heptomino; there are four of those that don't tile the plane. The one in Figure 2.27 is the most obviously bad in this context.

Figure 2.27: A heptomino that refuses to tile

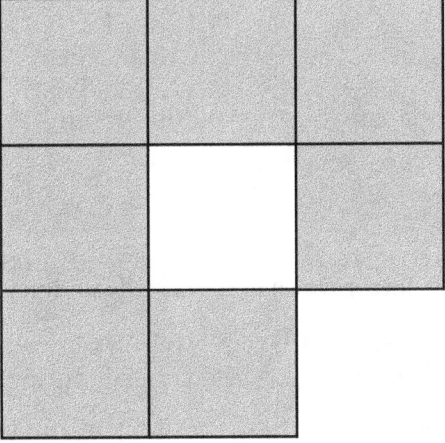

Chapter 5: Tessellations **87**

Level 4: Tiling rectangles with polyominoes

Problem

We now turn from the infinite to the finite. What *rectangles* can be tiled with dominoes?

Problem steps

Step 1

Let students experiment with concrete or virtual materials to see what they can find. The answer is any rectangle that has an area with an even value. To see this, note that the domino has an area of two (squares). If it tiles a rectangle, then the area of that rectangle must be divisible by two.

On the other hand, if a rectangle (R) has an even area, at least one of its sides must be even. Since we can tile a $1 \times$ even rectangle with dominoes, then we can tile R with dominoes.

These arguments proved this theorem:

> **Theorem:** A rectangle can be tiled with dominoes *if and only if* its area has an even value.

Step 2

Now ask your students what rectangles can be tiled with I-triominoes? Again it is experimenting time. It should soon look as if the following might be a good conjecture:

> **Conjecture:** A rectangle can be tiled with I-triominoes *if and only if* its area is divisible by three.

There needs to be an argument to justify this, which turns out to be similar to the domino theorem.

The I-triomino has three squares, so if it tiles a rectangle then that rectangle's area must be divisible by three. On the other hand, if a rectangle has an area divisible by three, then one of its sides is divisible by three. Since a $1 \times 3m$ rectangle can be tiled by m I-triominoes, the whole rectangle can.

So we get another theorem:

> **Theorem:** A rectangle can be tiled with I-triominoes *if and only if* its area is divisible by three.

Step 3

So what about I-tetrominoes: tetrominoes that have four squares in a row?

Perhaps the instant reaction is that they can tile any rectangle that has area divisible by four and no other rectangle. But what does the experimenting say?

Well, we fall flat on our faces at the first hurdle; you can't tile a 2×2 square with an I-tetromino! But maybe that is because the 2×2 square isn't big enough. So what about the 4×4 square? That is tileable.

How about the 2 × 6 square? No. What about the 6 × 6 square? Maybe that is big enough for the I-tetrominoes to wiggle around in.

Let students keep experimenting with I-tetrominoes and spaces until they get a feel for what might be possible, and better yet why it works out that way.

Step 4

Can you show that it is possible to tile a 4r × 4s rectangle with I-tetrominoes?

This is simple. The I-tetrominoes will tile a 4 × 1 rectangle, so they will also tile a 4r × 1 rectangle. Just put 4s of these together to tile the 4r × 4s rectangle.

Step 5

Can your students show that it *isn't* possible to tile a 6 × 6 rectangle with an I-tetromino? That is not so easy. We show the beginning of an argument in Figure 2.28.

Figure 2.28: The start of a proof that a 6 × 6 rectangle cannot be covered by the I-tetromino

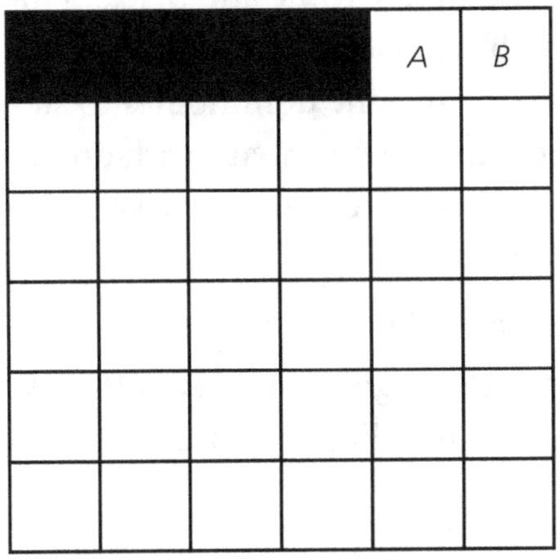

By symmetry, the only way to fill a corner square is to put an I-tetromino as shown in black. The only way to cover the A and B squares is to place two vertical I-tetrominoes as shown in Figure 2.29.

Figure 2.29: The next step in the proof

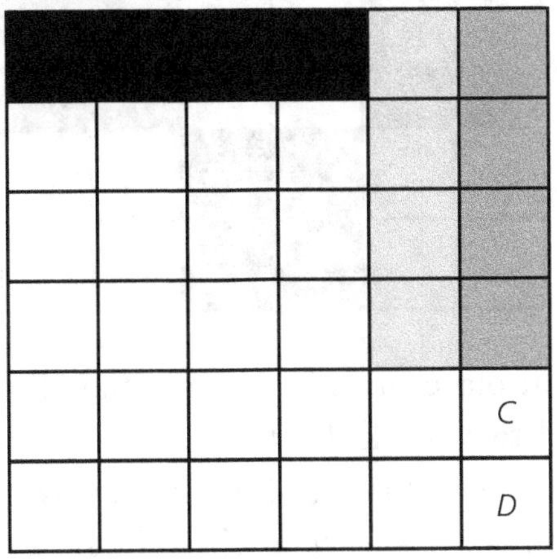

This forces the C and D squares to be covered by two horizontal I-tetrominoes. Continuing this argument around the space, we are forced to leave four squares in the middle of the 6 × 6 rectangle. These squares form a 2 × 2 square and so can't be covered by an I-tetromino.

Did your students have another idea?

Step 6

Can your students show that it isn't possible to tile a 66 × 66 rectangle with an I-tetromino?

We can try the method of the last step, but there are now too many cases to handle. It is not clear that we can even list them all, let alone deal with them.

So we take a side-track to a very old chestnut. Take a regular chessboard and remove the two diagonally opposite corner squares. Can your students tile the remaining board with dominoes?

This seems unlikely, but it is not easy to show until you remember that a chessboard has coloured squares. These are alternately white and black. Suppose the squares we removed from the chessboard are both black. Now think about a domino. When you try to tile the remaining squares of the chessboard, the domino has to sit on one white square and one black square. So if we are to tile the deleted board with dominoes, the board must have the same number of black and white squares. But it doesn't! Count them to see. So you can't tile the deleted board with dominoes.

This is the clue to the 66 × 66 rectangle problem.

We'll demonstrate the 6 × 6 rectangle problem another way and then leave the 66 × 66 rectangle to your students to sort out. Figure 2.30 shows the squares of the 6 × 6 rectangle coloured in 2 × 2 squares.

Figure 2.30: Colouring squares of a 6 × 6 array

Any I-tetromino that you place on this 6 × 6 rectangle will have to have two of its squares coloured white and two black. If we can tile the rectangle with I-tetrominoes, then two of its squares will cover white squares and two will cover black squares. So the I-tetrominoes will only cover the 6 × 6 rectangle if the rectangle has an equal number of white and black squares. And it doesn't!

Step 7

If r and s are odd, can you show that it is not possible to tile a $2r \times 2s$ rectangle with an I-tetromino?

Colour the rectangle with black squares and white squares using the same pattern used in the 6×6 rectangle on page 90. Then use the same argument that counts black and white squares to show that you can't tile the $2r \times 2s$ rectangle with I-tetrominoes if r and s are both odd.

This gives us a new theorem:

> **Theorem 4.1:** A rectangle can be tiled with I-tetrominoes *if and only if* one of the sides of the rectangle is divisible by 4.

Step 8

What rectangles can be covered with I-polyominoes that have n squares? Are there some values of n that are easier to deal with than others? Let your students play with these questions to see what conjectures they have. Encourage them to prove some of these conjectures.

It turns out that if n is prime or even, the answer to the tiling problem is essentially that of the two theorems in Steps 1 and 2 on page 88.

> **Theorem 4.2:** Let n be prime. A rectangle can be tiled with I-polyominoes with n squares *if and only if* one of the sides is divisible by n.

> **Theorem 4.3:** Let n be even. A rectangle can be tiled with I-polyominoes with n squares *if and only if* n divides the length of one of the sides.

The proofs are essentially the same as those in previous steps.

Where to from here?

- If n is composite, do we need n to divide one of the sides of the rectangle? What rectangles can we tile with I-polyominoes that have n squares? It might be worth experimenting with $n = 15$.
- What rectangles can be tiled with T-tetrominoes? This isn't easy, but may make a nice research topic. Students can look the problem up online and then write a report on what they have found.
- Suppose that we cut *any* two black squares from a chessboard. Can we cover the remaining board with dominoes?
- Suppose that we cut one black square and one white square from a chessboard. Can we cover the remaining board with dominoes?
- What ideas do your students have?

CHAPTER 6:
HOW HIGH IS A BUILDING?

Initial problem

How high is the highest point of your school? Is it building or a flagpole?

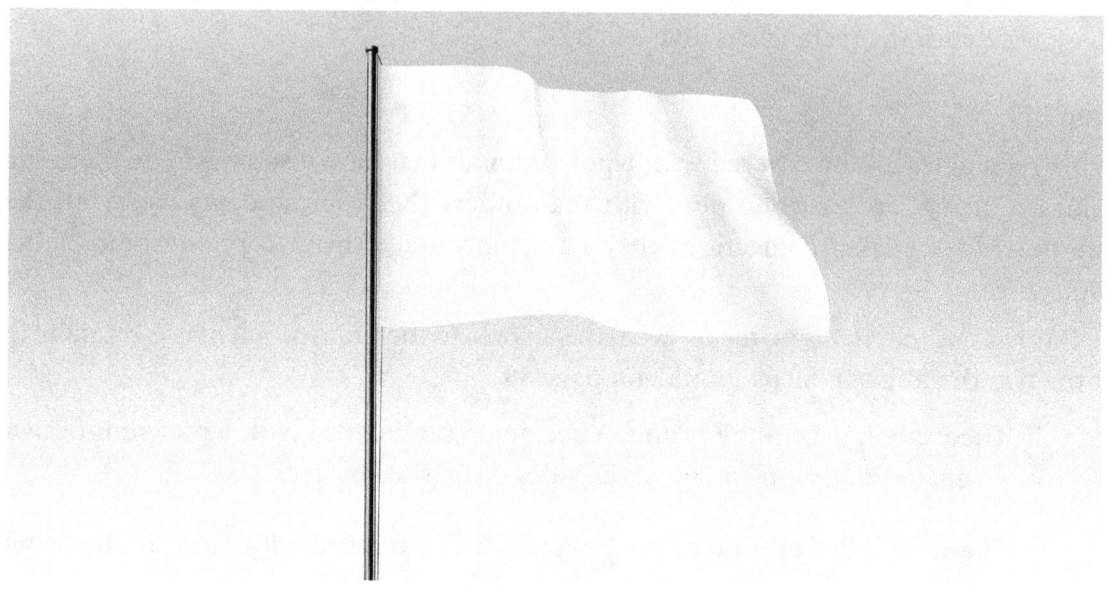

Background information

In this activity we assume that students can measure lengths greater than 10 metres and angles up to about 60°.

Throughout this activity we suggest the use of the following six pedagogical stages.

1. Estimate and record the height of the object (whole class).
2. Discuss possible ways to find the height and think about what measuring is required (whole class).
3. Plan how the measuring will be done (small groups).
4. Do the measuring (small groups).
5. Calculate the results (small groups) and think about possible errors.
6. Discuss the results including a consideration of errors and a check against the original predictions (small groups and then whole class).

The first stage is to help students get some idea of vertical distances. Students (and older people too) generally have a poor concept of vertical and horizontal distances.

The next stage may produce many different approaches to the problem. You may need to order these in priority or let different groups of students take different approaches. You may need to scaffold the class at this point in order for them to have a viable approach to the problem.

It is important to have a good idea of what is going to be measured, and by whom, before the students go out into the 'field'. Each group should consist of three or four members. This will enable two to hold each end of a tape measure, one to check the length and one to record the measurements.

The students should then make the required calculations in class. Each student should do these calculations so that they can be checked to be sure that there have been no errors made.

Finally, are the results reasonable? Do they make sense? (A flagpole, for example, is likely to be more than 100 cm high and less than 100 metres.) How do the measurements fit with the original predictions? If there is a significant difference, is this because the predictions were wrong, the measurements were wrong or the calculations were wrong?

It is worth listing all of the groups' results on the board because the class is likely to be amazed by the range of answers. So who is right and who is wrong? Clearly someone has made some errors, but where? At this point there is the need to look at where errors might have crept in.

In Level 1, without any knowledge of trigonometry, we find the height of a flagpole by using ratios associated with simple angles. Level 2 looks at measuring the same height using the shadow of a stick. Both of these levels are accessible to students in Year 7 and above.

Students construct an inclinometer in Level 3 and measure the flagpole again. This level is aimed at Year 8 students.

Finally, trigonometry is used to measure heights in Level 4. Some Year 8 students will be able to tackle this level, but it is designed for Year 9 students.

Table 2.4: Australian Curriculum content descriptions for the *How high is a building?* activity

Activity level	Problem	Content descriptions
1	The flagpole	*Year 7* Classify triangles according to their side and angle properties and describe quadrilaterals (ACMMG165) Recognise and solve problems involving simple ratios (ACMNA173)
2	Sunny days	*Year 7* ACMNA173 (see above) Describe and interpret data displays using median, mean and range (ACMSP172) *Year 8* Solve a range of problems involving rates and ratios, with and without digital technologies (ACMNA188) *Year 9* Solve problems using ratio and scale factors in similar figures (ACMMG221)

Table 2.4: Australian Curriculum content descriptions for the *How high is a building?* activity (continued)

Activity level	Problem	Content descriptions
3	Inclinometers	*Year 8* Define congruence of plane shapes using transformations (ACMMG200) Establish properties of quadrilaterals using congruent triangles and angle properties, and solve related numerical problems using reasoning (ACMMG202)
4	Using trigonometry	*Year 9* ACMMG221 (p. 93) Use similarity to investigate the constancy of the sine, cosine and tangent ratios for a given angle in right-angled triangles (ACMMG223) *Year 10* Apply logical reasoning, including the use of congruence and similarity, to proofs and numerical exercises involving plane shapes (ACMMG244) Solve right-angled triangle problems including those involving direction and angles of elevation and depression (ACMMG245)

Problem aims

» Using and constructing simple geometrical instruments to find heights
» Finding heights without and with trigonometry

Key concepts

» Ratios
» Angles
» Trigonometry: sin, cos and tan

Possible heuristics/strategies

» Trial and error
» Break a problem into manageable parts

Concrete materials

» Sticks
» Tools to measure up to 20 metres
» Protractors
» Weights for inclinometer

Special note

Inclinometer: An instrument for measuring angles of elevation and depression.

Level 1: The flagpole

Problem

How high is the school flagpole?

Problem steps

In this level we assume that students don't know any trigonometry.

It would be good, but not necessary, if you knew the exact height of the school flagpole before students started trying to find its height. This would give everyone a good idea of how accurate their measurements and calculations are.

We have largely ignored the fact that most angle measurements will be made by a standing student. This means that adjustments have to be made because of the height of the measurer. We leave it to the class to make the appropriate adjustments.

Step 1

There are certain problems with measuring the height of a building, so this activity starts by first looking at something simpler. We have chosen a flagpole but you could use football goalposts, the height of a classroom's wall or many other options; the principles are the same. The object only needs to have a clear line of sight with nothing around it and be more or less perpendicular to the ground.

There is more than one way to measure a flagpole's height. One way is to measure the rope that pulls up the flag. If you are careful this can be done without taking the rope off the pole. But you may have to add extra bits (and errors) because of things like the distance from the top of the rope to the top of the pole and the distance from the ground to the bottom of the rope.

What other methods can your students think of? How well do they work? What heights do they get? (No, they are not allowed to chop it down.)

Step 2

How can we find the height of the pole less directly?

Introduce the idea of a triangle with the flagpole being one side. Get them to imagine the situation. T is the top of the flagpole and B is the bottom, but where to place point G for Ground? What can your students tell you about the triangle GTB?

Figure 2.31: The flagpole

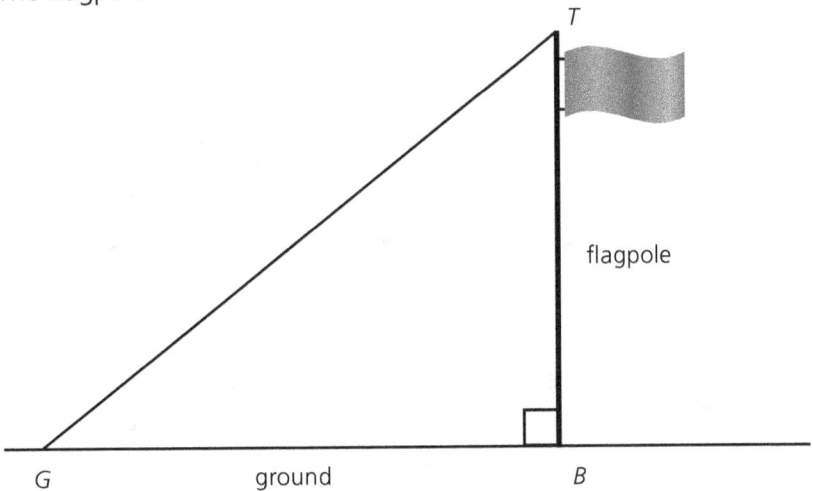

Well, G could be anywhere and GTB is a right-angled triangle provided the ground is horizontal. But how does this help? What is the angle at G? It could be any angle you wanted it to be (provided it is less than 90°). What angle would you like it to be and why?

We could posit 45° as a good angle. That makes triangle GTB an isosceles triangle and so GB = TB. This means that you only have to find some point G on the ground so that angle BGT is 45°, measure GB and you know the height of the flagpole. Is there such a point? How would you know if you found it?

It looks as if some careful use of a protractor is needed. Get the groups to work out how they are going to do this. Then let them measure GB.

What answers did they get? Where did the errors come in? They will probably sight the top of the pole with a protractor and standing up. They need to remember to add their height into the calculation. There will almost certainly be an error in measuring GB, but the biggest error will come in the measuring of 45°. How can they minimise this error? How can they hold the protractor steady and horizontal?

Step 3

What other angles might be used at point G?

Let the students choose something relatively easy to measure. How about 60°? At this point they may not realise that TB/BG is a fixed ratio. Help their intuition by getting them to accurately draw three right-angled triangles with a vertex of 60°. Make sure that they make things easier for themselves by using whole numbers for BG. Get them to measure TB and BG in each case.

We give some possibilities in Figure 2.32.

Figure 2.32: Some right-angled triangles with an angle of 60°

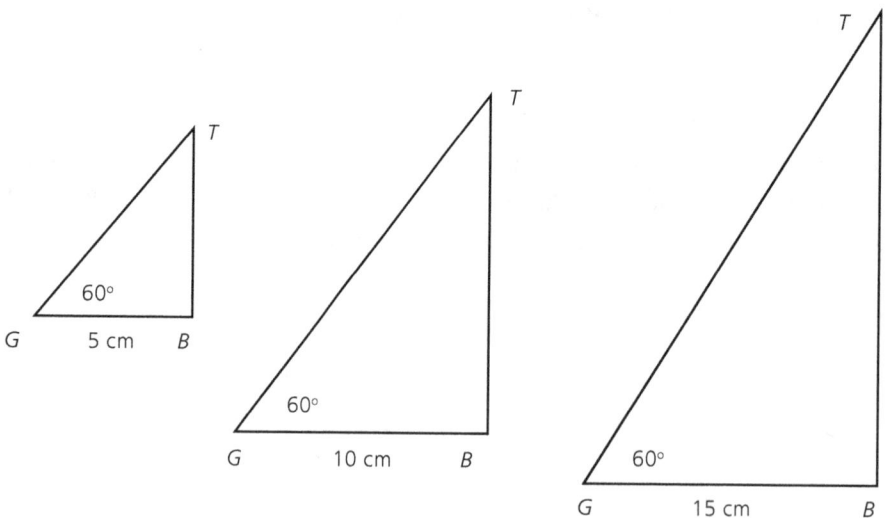

Get the class to put their results for BG and TB on the board. Can they see any patterns in the TB measurements? Hopefully it looks as if TB changes but TB/BG stays the same at about 1.7.

What do they predict that the value of TB would be if BG = 20 cm?

It turns out that TB/BG is very close to 1.7. Actually it is closer to 1.732 and it is precisely equal to $\sqrt{3}$. This comes about because the three triangles in Figure 2.32 are *similar triangles*—triangles with all angles the same.

Ask the students if they know another angle that makes *TB/BG* constant no matter what *BG* is. (They have seen 45° earlier, but they may be able to predict that it is true for all angles.)

Step 4

It is time to go out into the school grounds and for each group to find three *G*-points where the angle to the top of the flagpole is 60°. Make sure that they measure *BG* too. Then come back into the class to discuss their results.

Step 5

Some things to think about now:

- Are their answers using 60° significantly different to the answers they worked out using 45° before?
- Are 45° and 60° the only angles for which *TB/BG* is a constant?

Where to from here?

- Is there anyone in the school who knows precisely how high the flagpole is? If so, the students will have a check on their answers.
- What heights around the school *cannot* be measured this way? Why?
- How would a surveyor measure the height of the flagpole?

Level 2: Sunny day

Problem

How can you find the height of a flagpole on a sunny day?

Problem steps

This is the same problem as in Level 1 but we try to do it another way—using the sunshine.

Step 1

Remind the students of the situation in Figure 2.31. This time, assume that the sun is shining and G is the end of the flagpole's shadow. Ask them how they might find the height of the flagpole using shadows. Can they use other shadows? Is there anything else whose shadow ends at the same point as the flagpole's? Can we find such an object? How would that help?

Look at the diagram in Figure 2.33.

Figure 2.33: The flagpole and the object

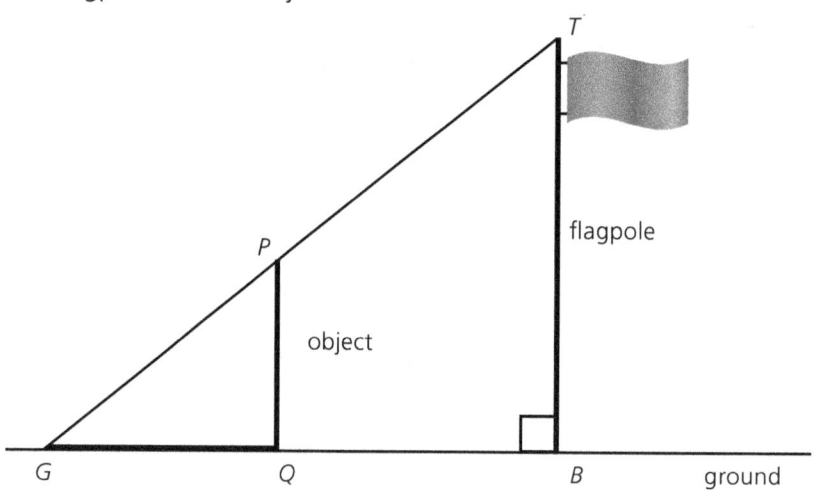

Suppose that we put a stick PQ between B and G so that G is both the end of the pole's shadow *and* the end of the stick's shadow. From what we learned in Level 1, it is likely that the ratio of $TB : BG$ is the same as the ratio of $PQ : QG$. This is in fact true by a property of similar triangles (triangles GBT and GQP). So:

$$TB = \frac{(PQ \times BG)}{QG}$$

By placing a stick in a suitable position and measuring PQ, BG and QG, it should be possible to find the height of the pole.

Give students the opportunity to find the height of the pole this way.

Step 2

What if the sun isn't shining? Students may see that if they lie on the ground and instruct someone to move the stick so that they see the top of the stick and the top of the pole aligned, the point where they are lying could be thought of as the end of the shadow—and now they are back in Step 1 again. Let them try this method out for measuring the height of the flagpole.

What errors can they think of with this measurement?

Which method for finding the height of the pole is likely to be the most accurate? (Any time you use angles there is going to be an error. The two problems here are getting the horizontal to measure from and reading the protractor accurately. This latter might be made easier by magnifying the size of the protractor, but how can your students fix the horizontal problem?)

Step 3

By now a reasonable amount of data has been collected. It is worth the students graphing this as a dot plot with dots, crosses and so on for the various methods of measurement from Steps 1 and 2.

What height does this suggest for the flagpole? If there are any differences, why does the class think these have happened? Is one method better than the others?

Step 4

Calculate the mean heights for the two measurements. Do the two means differ by much? Why do students think that they do—or don't? If there is a difference, which do they think is the more likely to be correct and why? What does the class think that the mean of the measurements is?

It would be good at this point to confirm the actual height of the pole if it is known, or get it measured accurately if you have a friendly surveyor on hand.

Step 5

It might be worth the class using the different measuring techniques on something more accessible and where the height can be measured more directly, such as the height of a netball ring.

Where to from here?

- What other ways are there of measuring heights?
- How could the class measure the height of something where the ground is not level?
- What if it is not possible to measure the length *GB* directly because there are buildings surrounding the object *TB*?

Level 3: Inclinometers

Problem

One of the major errors in the measurements so far has been finding the correct angle. How can this error be reduced?

Problem steps

Step 1

Discuss with the class how they might improve their angle measuring. Get them to design and produce a better method. They could then use their invention to measure angles in the field.

Hold a class discussion on the benefits or otherwise of their designs.

Step 2

What the class has been trying to construct in Level 1 is an *inclinometer* (or *clinometer*). Let them investigate for these devices online and build one from a set of instructions. (You can find links to such instructions via the series website.)

Step 3

Let the class use the inclinometer in the field. How does it compare with the protractor method used in Levels 1 and 2?

What do they now think the height of the flagpole is? Does this vary much from the answers found in Levels 1 and 2? Why do they think this is the case? Is this simply the use of the inclinometer?

Where to from here?

I Your students could do some research on surveyors and try to answer the following questions: What instruments does a surveyor use? What is each one used for? How are they used? Can you get a surveyor to talk to the class and demonstrate their instruments?

Level 4: Using trigonometry

Problem

How high is the highest point of your school building?

Problem steps

Step 1

First, return to the flagpole of Level 1. Remind students that the ratio *TB/GB* is fixed for different values of the angle *BGT*. Tell them that the ratio *TB/GB* is called the *tangent* (tan) of the angle *BGT*. The values of the tangent have been worked out for all possible angles; it is just a matter of pushing a button on their calculator.

Now send them out into the field to measure a variety of angles of *BGT* (with different positions of *G*) and the corresponding values of *GB* for the flagpole. (It is a good idea for them to measure each angle three times with their inclinometer and take an average of the readings later.)

Once they have finished measuring, they can produce their calculations in class. Groups should graph their answers, delete any outliers and produce a mean of their calculations; the whole class can then determine a class mean.

How close to the precise height of the pole did they get?

Step 2

Now ask the students to find the height of the tallest building at school. It is likely that they can't do this using the established methods because the tallest point won't be on the edge of a building. So the highest point will need to be sighted past other buildings and *BG* can't be found directly.

Get your students to think about how they can now measure this height. We show two methods in the next two steps.

Step 3

Method I: Assuming that there is a direct line of sight to the top of the building, measure this direct line of sight, *GT*.

But how can you do this? This might be accomplished using a single lens reflex camera. They focus automatically, and some lenses tell you how far away things are when they are focused because they have an on-board range finder. If such a camera isn't available, students might use one of the range finder apps available for smartphones and tablets. They could even make one - online instructions for making a simple range finder can be accessed via the series website.

This method requires the class knowing the *sine* (sin) of an angle, because *TB/GB* = sin (angle *BGT*).

Step 4

Method II: For this method, your students need to take some more measurements before they get the length *GT*. We show this in Figure 2.34.

Figure 2.34: Measuring an inaccessible height

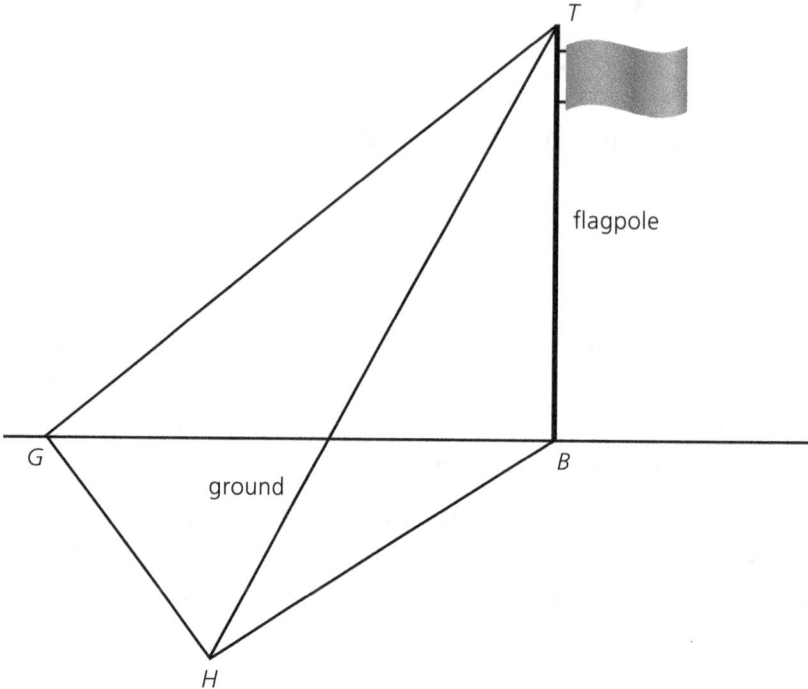

Here your students will need to measure *GH* and angles *GHT*, *HGT* and *BGT*. It will be best if they can make sure that angle *BGH* = 90°, because this will ensure that angle *HGT* is also 90°. (Why?)

But how can the class measure angle *GHT*? This isn't easy; it is the first angle they have had to measure that is not in the vertical plane, so the inclinometer will not be as useful as before. Students will need to sight *T* from *H* so that the protractor is in the plane *GHT*, then turn the protractor until the horizontal line on the protractor points at *T*. This is most easily done if they have something to rest the protractor on.

Given that triangle *HGT* is a right-angled triangle, and that they have measured *GH* and angle *GHT*, they can calculate *GT* in the usual way.

Now they have also measured angle *BGT*, so *BT*/*GT* = sin (angle *BGT*). Hence they can find *BT*, the height of the building.

Where to from here?

- What if the land is not horizontal between where they have stood at *G* and *H* or *B*? How do they need to adjust their calculations?
- Can they use this method to find the height of a mountain? How do surveyors find the heights of mountains?
- Why did the first surveyor who found the height of Mount Everest lie about his calculations? (They were too close to a round number and he thought people would say he had rigged it somehow.)

PART 3: STATISTICS AND PROBABILITY

Part 3 presents three activities centred on the Statistics and Probability strand.

Table 3.1: Statistics and Probability exercises

Problem	Big ideas
Greedy Pig	- Addition of small numbers - Collecting and recording data - Using properties of numbers to continue patterns - Generalising from number properties and results of calculations
Pascal's triangle	- Using properties of numbers to verify patterns in Pascal's triangle - Discovering binomial coefficients - Using binomial coefficients to find the size of sets - Using binomial coefficients to determine probabilities
Monty Hall's problem	- Probability in real situations - The probability of two situations occurring together or not

Some reminders before you use these tasks in your classroom:

1. The questions in the text are ones you can ask your students. You are likely to be able to produce similar, more immediately relevant ones for your particular students as you work on these activities with them.

2. We have given suggested links to the Years in the Australian Curriculum: Mathematics for all the Levels in each activity but, given that there will be a spread of ability in your classes, you should take these as a guide only. Take the opportunity to encourage every student to the edge of their comfort zone.

3. To take all students further, sometimes you can omit some of the later steps of a Level in favour of the early steps in the following Level.

CHAPTER 7: GREEDY PIG

Initial problem

Xi and Maeve play a game where they each roll a six-sided die. In each round they either stop rolling when they roll a 1 or when they decide they need to stop. If their die shows 1, their score for that round is zero; if their die comes up with any other result, they add that number to what they have got so far in the round. At the end of each round, each player adds that round's score to what they have got so far from other rounds. The first one to get a score of 50 wins.

What advice would you give to Xi and Maeve about playing the game?

Background information

There is no clear history of *Greedy Pig*, but it and similar games are well-known internationally. It is valuable in class as it gives students the opportunity to play a dice game that incorporates some useful techniques including stem-and-leaf diagrams, box plots and tree diagrams, while collecting data for the purpose of discovering good strategies to win.

In Level 1 we simply play the game and get a feel for it. In the process, students begin to see a number of conjectures that might make valuable strategies. This level is open to all students as it is simply about playing a game, collecting data and making conjectures about outcomes.

Level 2 progresses to collecting data that will enable students to make conjectures for good strategies. Level 2 is also open to students from Year 7 onwards, who can use the data collected to test their conjectures from Level 1.

At Level 3, students use their data to play off their conjectures against each other. While most of this activity is still accessible to Year 7 students, some elements are more appropriate for Year 8 and Year 9 students.

In the final Level we give a semi-theoretical basis for two kinds of conjecture. The theoretical consideration of conjectures at this level lifts it above the ability of most students under Year 9.

Table 3.2: Australian Curriculum content descriptions for the *Greedy Pig* activity

Activity level	Problem	Content descriptions
1	Simulation for conjectures	*Year 7* Identify and investigate issues involving numerical data collected from primary and secondary sources (ACMSP169) *Year 8* Investigate the effect of individual data values, including outliers, on the mean and median (ACMSP207)
2	Simulation for verification	*Year 7* ACMSP169 (see above) Construct and compare a range of data displays including stem-and-leaf plots and dot plots (ACMSP170) Calculate mean, median, mode and range for sets of data. Interpret these statistics in the context of data (ACMSP171) Describe and interpret data displays using median, mean and range (ACMSP172) *Year 8* ACMSP207 (see above)
3	The battle of the conjectures	*Year 7* ACMSP169 (see above) ACMSP170 (see above) ACMSP172 (see above) *Year 8* ACMSP207 (see above) *Year 9* Construct back-to-back stem-and-leaf plots and histograms and describe data, using terms including 'skewed', 'symmetric' and 'bimodal' (ACMSP282)
4	Proof	*Year 9* List all outcomes for two-step chance experiments, both with and without replacement using tree diagrams or arrays. Assign probabilities to outcomes and determine probabilities for events (ACMSP225) ACMSP282 (see above)

Big ideas

- » Simulation
- » Recording data (stem and leaf, box and whiskers, tree diagram)
- » Making conclusions from data

Problem aims

- » To put probability knowledge to use in a well-known game
- » To see the value of simulation

Key concepts

- » Simulation
- » Tree diagrams

Possible heuristics/strategies

- » Simulation
- » Look for patterns
- » Conjecture, check and prove

Concrete materials

- » Dice (enough so that each student has one)

Level 1: Simulation for conjectures

Problem

Xi and Maeve play a game where they each roll a six-sided die. In each round they either stop rolling when they roll a 1 or when they decide they need to stop. If their die shows 1, their score for that round is zero; if their die comes up with any other result, they add that number to what they have got so far in the round. At the end of each round, each player adds that round's score to what they have got so far from other rounds. The first one to get a score of 50 wins.

What advice would you give to Xi and Maeve about playing the game?

Problem steps

Step 1

Describe the game and the question to the class. Make sure that they understand the rules, then let them play the game by themselves. Ask them what they think might be the best way to get to 50 in the fewest number of rounds.

Give them 10 minutes to do as many trials as possible. Tell them you will then have a class discussion about the best strategies. Don't say anything about recording data.

Step 2

Have a general discussion to see what they have decided from their experiments. Let them propose a winning strategy and say why they think it is a good one. Other students may want to support or reject the strategies of the others. If the strategy looks as if it might be good, write it on the board.

There are likely to be a range of ideas suggested, such as those listed below.

- I stopped after three rolls on each round because I was worried that the number 1 would come up.
- I stopped when I got 25 as then I only needed two rounds to get 50.
- I just rolled the die and stopped when I felt like it.
- The number 1 comes up one time in six, so I thought it might be best to roll five times before I stopped.

Step 3

Get each student to choose a strategy and write it down.

Now play a game against the entire class. Roll your die at the front of the class, announcing your results, while each student plays according to the strategy they have chosen. When the students decide to stop, they must write their score for each round and their total so far. They must keep quiet except when they reach 50, when they can call out 'Greedy Pig'.

After a couple of games, have another discussion to see if the students think the strategy they chose was a good one or whether there might be a better one. They need to support their claims with some argument.

Step 4

Repeat Step 3, but any student can change their strategy. However, they have to stick with this new strategy throughout the game.

You may want to repeat Step 3 a few more times until you think that their intuition has developed enough to move on.

Step 5

Now consider the strategies that were written on the board in Step 1. Which of these don't look so good now? Which should be kept? Are there any new strategies that they wish to add?

Label these as conjectures. You might want to call them Conjecture 1, Conjecture 2 and so on, or name each after the students who suggested them.

Where to from here?

- What do your students suggest is the next move with this game?
- Can they think of any extensions or variations? For example, they might give everyone a free roll or two at the start of a round where there is no penalty for getting a 1. They might also suggest using two dice or even coins in various ways.

Level 2: Simulation for verification

Problem
How can we test our conjectures about strategies for playing Greedy Pig?

Problem steps

Step 1
The aim here is to systematically collect more data to lend support to some conjectures, as well as to make others seem less reasonable and so abandon them. If each student can collect data for five rounds, you might have 100 sets of game data to help the investigation.

The key thing is really the round data. After all, Xi and Maeve only have control over what they do in each round. So the round data is more important and this depends on the rolls. It is therefore worth concentrating on rolls and rounds.

The data has to be carefully recorded if it is to be used in any meaningful way. So the other issue here is what data to record and how to record it.

Discuss the importance of the rolls and rounds with the class, as well as what data to keep and how to record it. The following pieces of information are the most important to collect, as these are the basis of most conjectures. Ask the students to identify:

1. the number of times the numbers 2 to 6 occur and their mean
2. the number of rolls before a 1 comes up and the mean of that number
3. the score obtained in a round
4. the score obtained after a given number of rolls in a round.

It would be useful to use box plots for points 1 and 2 and stem-and-leaf plots for points 3 and 4.

Step 2
Get the class to collect the data in small groups. Then suggest they do the statistical analysis of each set of data. How do these help the conjectures proposed in Level 1?

Return to the points outlined above.

1. This data gives some idea of what score they will get on average for each roll of the die.
2. This data suggests how many rolls are likely before they will get a 1.
3. This data will help students get a feel for which total comes up most often.
4. This data suggests how many rolls they will most often make before they get a 1.

The results of this experimenting may be limited due to the number of rolls or trials students can make in a given lesson. There are a variety of die-rolling programs online (see the links on the series website) that can produce a lot more rolls and hence more accurate long-term data. Check the computer results against the class results.

Step 3

Let the class discuss in groups the significance of these results. What conjectures seem the best now?

Discuss this with the whole class.

Where to from here?

- Do the results support all the conjectures that your class has made? Can the ones that are not supported be changed slightly because of the data to make better conjectures?
- How does the data help to find strategies for the game variations suggested by your students in Level 1?

Level 3: The battle of the conjectures

Problem

Which of the remaining conjectures gives the best strategy for playing Greedy Pig?

Problem steps

Step 1

Discuss what conjectures are left after the experimentation of Level 2.

It is likely that the conjectures will have been reduced to two at this point.

Conjecture 1: A winning strategy is to roll five times in a round and then stop.

Conjecture 2: A winning strategy is to keep rolling until a total of 20 is reached and then stop.

However, it is possible that other strategies have been supported by the data. For example, a variation of the first conjecture might be:

Conjecture 1A: A winning strategy is to roll six times in a round and then stop.

While a small variation of the second conjecture could be:

Conjecture 2A: A winning strategy is to keep rolling while the total for the round is less than 20. Once this total is reached, stop rolling.

Some students might even want to combine the two to give:

Conjecture 3: A winning strategy is to keep rolling for five rolls or while the total is less than 20, whichever comes first.

Consider the situation in a class discussion. Get students to vote on their favourite conjecture. The discussion should include supporting arguments.

Step 2

Divide the class into teams based on which conjecture they support, then stage a 'shoot-out'. In small groups, members of one team play their strategy against the supporters of another strategy. But first, the class needs to determine a reasonable number of games (first to 50) to play to determine the winning conjecture.

Step 3

Have the groups report back to the class as a whole and discuss the results of the shoot-out.

What is now considered the best conjecture, and why?

Step 4

The 'why' is important. Can the class find theoretical reasons for their choice of conjectures? (We will give this in Level 4.) Conjecture 1 may look good because you would expect to roll a number other than 1 five times out of six—so it might be worth rolling five times and hoping that your luck lasts.

Is there a similar reason for the 'about 20' strategy of Conjecture 2?

Where to from here?

- Recall the variations of *Greedy Pig* from previous steps. Analyse these to see what a best strategy might be.
- Have the students found a better way to represent the data than with stem-and-leaf plots or box plots? What kinds of data do they think are best represented by stem-and-leaf or box plots? Why?

Level 4: Proof

Problem

How can we prove that our Greedy Pig strategies actually work?

Problem steps

Step 1

In this level we give a strong reason for each of Conjectures 1 and 2. First, we could look at Conjecture 1.

Go back to the mean of the rolls of 2, 3, 4, 5, and 6 (see Point 1 and the discussion in Steps 1 and 2 of Level 2). The mean that was found there was probably close to 4. You might argue this by saying that on average you would expect $2 + 3 + 4 + 5 + 6 = 20$ in 5 rolls, so the average (mean) is $\frac{20}{5} = 4$. On average, then you might hope for a total of 4 for a roll. So no matter what happened on the last roll, the next one might be considered to be 0 (you rolled a 1) or 4.

This can be represented in the tree diagram in Figure 3.1.

Figure 3.1: The possibilities on a single roll

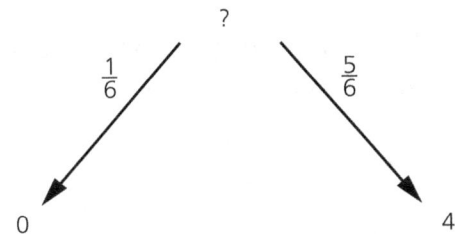

Talk your students through this diagram.

Step 2

Now move on to the bigger picture. Ask the class to draw the complete tree diagram of a round.

What happened in Figure 3.1 happens time and time again as we develop the fuller tree diagram of Figure 3.2.

Figure 3.2: The infinite tree diagram for Greedy Pig

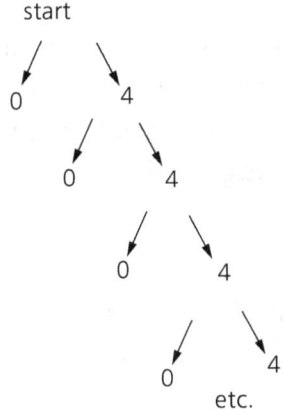

So what would a player expect to gain in the game? After one roll it would be $4(\frac{5}{6})$, as she would get a total of 4 five times out of six. After two rolls she would expect to get $8(\frac{5}{6})^2$, as her total would be 8 but the chances of getting there would be $(\frac{5}{6})^2$. After three rolls she would expect $12(\frac{5}{6})^3$ and so on. In general, this gives $4n(\frac{5}{6})^n$ after n rolls.

Step 3

If this is the probable result, when would the player want to stop? This might be when her hoped-for gain stopped increasing; at this point it is not worth her while to continue. So ask the class: when does the gain start to diminish?

They can do this by calculating different values of $4n(\frac{5}{6})^n$ starting with $n = 1$ and working on up. Alternatively they can find out when $4n(\frac{5}{6})^n$ is greater than $4(n+1)(\frac{5}{6})^{n+1}$.

Now, $4n(\frac{5}{6})^n > 4(n+1)(\frac{5}{6})^{n+1}$ implies that $n > (n+1)(\frac{5}{6})$, and this happens when $n > 5$.

Step 4

But what if $n = 6$? (We at least had Conjecture 1′ that $n = 6$ was best.)

Well, when $n = 5$, $4n(\frac{5}{6})^n = 8.03755$. But that is exactly the same result as when $n = 6$! Check it out. So that gives some justification for Conjectures 1 and 1′.

Step 5

But where do Conjectures 2 and 2′ come into the picture? Ask the class how they can use a tree diagram to get this result.

Another approach here is to look at when our contestant's gain starts to go down. Figure 3.3 is another way of looking at the tree diagram situation.

Figure 3.3: The possibilities on a single roll

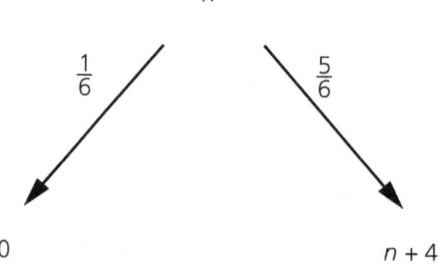

If the contestant has built up a score of n, she will lose n if she rolls a 1, but gains 4 otherwise.

So her net gain is 4 with a probability of $\frac{5}{6}$, or $-n$ with a probability of $\frac{1}{6}$, and is therefore $4(\frac{5}{6}) - n(\frac{1}{6})$.

The class can calculate when this is negative. It turns out to be when $n > 20$. So if she has a total of less than 20, she should roll again; if she has a total of greater than 20, she should stop.

What should she do if her total is exactly 20? (Her gain is 0 so she can do what she likes, though it is probably not worth taking the risk.)

Where to from here?

1. What if the game allows a free turn for the first roll in each round? In other words, the contestant can keep a total of 1 if she rolls 1 on the first roll. On the other hand if she gets something other than 1 she keeps that too. What are her best strategies? Get the class to analyse this situation and the situations of any other games that are like *Greedy Pig*.

CHAPTER 8: PASCAL'S TRIANGLE

Initial problem

What do your students think that the next row of this triangle of numbers is? Why? What other patterns can they find here?

```
              1
            1   1
          1   2   1
        1   3   3   1
      1   4   6   4   1
    1   5  10  10   5   1
  1   6  15  20  15   6   1
```

Background information

This activity is based around Pascal's triangle and its value in counting and finding probabilities.

During his short life (1623–1662), Blaise Pascal made significant contributions to both mathematics and philosophy. He was one of the first people to make a mechanical calculating machine and he developed the mathematical theory of probability with Pierre de Fermat.

Pascal's triangle is a particularly nice array of numbers where the rows are connected by an addition property (the sum of two adjacent numbers on one row add to give the number between them on the row below). There are many patterns to be found in the triangle, including a diagonal consisting solely of Fibonacci numbers, so it is interesting to investigate in its own right. However, it has important practical applications because all of the entries of the triangle are binomial coefficients and therefore are tied to both the expansion of $(x + 1)^n$ and to counting and probability.

It is worth noting that the triangle was not invented by Pascal; it was known for centuries before Pascal took an interest in it. And the triangle is not universally given his name; for example, it is known as Tartaglia's triangle in Italy.

»

Level 1 introduces Pascal's triangle and encourages students to look for patterns in the numbers, one of which is the 'hockey stick' result. Level 1, except possibly Step 6, should be accessible to all Year 7 students.

In Level 2 we look for more patterns and show the relation with binomial expansions and binomial coefficients. Level 2 might cause problems for Year 8 students unless they have already done a reasonable amount of algebra.

The Fibonacci numbers in Level 3 can be seen as a pattern to be analysed. Year 7 students can do a lot of Level 3, and Year 9 students should be able to do all of it. There are no content descriptions in Table 3.3 because the mathematics required from the content is nothing more than addition. On the other hand, using this knowledge requires a certain maturity on behalf of the students.

Counting using binomial coefficients leads to finding probabilities in Level 4, which is for Year 9 students.

Table 3.3: Australian Curriculum content descriptions for the *Pascal's triangle* activity

Activity level	Problem	Content descriptions
1	Pascal's triangle	*Year 6* Identify and describe properties of prime, composite, square and triangular numbers (ACMNA122) *Year 7* Assign probabilities to the outcomes of events and determine probabilities for events (ACMSP168)
2	Other patterns	*Year 7* ACMSP168 (see above) Create algebraic expressions and evaluate them by substituting a given value for each variable (ACMNA176) *Year 8* Factorise algebraic expressions by identifying numerical factors (ACMNA191)
3	Fibonacci's sequence	
4	Getting in a team	*Year 8* Identify complementary events and use the sum of probabilities to solve problems (ACMSP204) *Year 9* Extend and apply the index laws to variables, using positive integer indices and the zero index (ACMNA212)

Big ideas
» Binomial coefficients
» Counting and probability

Problem aims
- Get to know Pascal's triangle and its properties
- Find patterns in Pascal's triangle
- See Fibonacci numbers and some of their properties
- Apply Pascal's triangle to counting and probability

Key concepts
- Binomial coefficients
- Fibonacci numbers
- Counting without directly counting

Possible heuristics/strategies
- Try simple cases
- Look for patterns
- Guess and check (trial and error)

Level 1: Pascal's triangle

Problem

What do your students think that the next row of this triangle of numbers is? Why? What other patterns can they find here?

Figure 3.4: Pascal's triangle

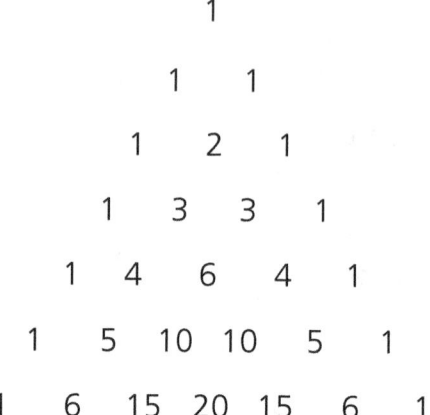

Problem steps

Step 1

Pascal's triangle is a mathematical construct that everyone should know about. If nothing else, it has lots of interesting patterns. What is the next row? What other interesting patterns are there?

On their own or in groups, give your students a chance to have a look at Pascal's triangle and think about what they see. After a while, bring them together to discuss their thoughts. We'll concentrate first on how they get the next row. Other patterns they might find will be discussed in Step 6 and Levels 2 and 3.

Every row begins and ends with 1. The second and second-last numbers are the number of the row. As the top of the triangle has no second number, we'll call this *Row zero*. So the next row in the partial triangle above would be the seventh. It should look like 1 7 … 7 1. If this pattern continues, then the nth row is 1 n … n 1.

In the fourth row we get more than four numbers for the first time. The extra number here is 6. Where does that come from? Is it an accident that the two numbers above it in the third row are 3 and 3, and 3 + 3 = 6? No. Look at the rows that come after the fourth row. Apart from the diagonals of ones, *every* number is the sum of the two numbers above it. This means that we can probably construct more and more rows of the triangle, giving it an infinite number of rows.

Step 2

Do the numbers in Pascal's triangle have any significance? Discuss this with your class and then let them find the answers to the following questions.

Suppose you had three friends, Alain, Bauji and Kapil.

- In how many ways can you invite *none* of them to come over to play a videogame? (Just one—you don't invite any of them.)
- In how many ways can you invite *one* of them to come over to play a videogame? (There are three possibilities; just Alain, just Bauji or just Kapil.)
- In how many ways can you invite *two* of them to come over to play a videogame? (Three again—Alain and Bauji, Alain and Kapil or Bauji and Kapil.)
- In how many ways can you invite *three* of them to come over to play a video game? (Just one—you invite them all.)

Now the numbers 1, 3, 3 and 1 might look pretty familiar. They are the numbers in row three of Pascal's triangle.

Step 3

What if someone in your class has six friends? In how many ways could they invite 0, 1, 2, 3, 4, 5 or 6 of them to some event?

Let the students work on this in small groups. When they come back to discuss the answers they should have found 1, 6, 15, 20, 15, 6 and 1. This looks like the sixth row of Pascal's triangle. Why?

Step 4

If a student has 10 friends and wants to choose four of them for some activity, how many choices are there? Can the students determine this without writing out all of the possibilities? Get them to think about an efficient way to do this.

You can choose the first one in 10 ways, the second one in 9 ways, the third one in 8 ways and the fourth one in 7 ways. So it might look as if there are $10 \times 9 \times 8 \times 7 = 5040$ ways to do the selecting. But why is this not true?

Suppose you chose A, B, C and D. You might have chosen A first, then B, then C, then D. But you might also have chosen A, then C, then B, then D. How many ways, then, can you order four people? There are four possibilities for the first choice, three for the second, two for the third and one for the fourth. This is $4 \times 3 \times 2 \times 1 = 24$ options.

The number of choices of four friends from 10 is therefore $\frac{5040}{24} = 210$. Is this number anywhere in the 10th row of Pascal's triangle?

Step 5

It turns out that the *r*th position in the *n*th row of Pascal's triangle is the number of ways of choosing *r* things out of *n*. So the entries in the triangle relate to some counting situations. The formula that gives these numbers is:

$$\frac{n(n-1)(n-2)\ldots(n-r+1)}{(r \times r-1 \times \ldots \times 3 \times 2 \times 1)}$$

We'll call this ^nC_r and it is the number of ways of choosing r things from a set of n things.

Step 6

In the picture in Figure 3.5, the italic numbers all add up to the bold number. Ask the class if this always the case, no matter which diagonal numbers you take to add together.

Figure 3.5: Adding diagonal numbers in Pascal's triangle

(No, it is not. They should see that they have to take the diagonal numbers starting with a '1'.)

Did any of your students come up with this pattern? If so, they may have already discovered the 'hockey stick (or Christmas stocking) theorem', named after the shape that the numbers form. The 'toe' number is the sum of the others.

Can your class justify this? It is a direct result of the way the triangle is made up. They can look online to find proofs of this, including animated proofs. (See the series website for links.)

Where to from here?

- What other things can your class see in Pascal's triangle that we haven't mentioned so far?
- Experiment with your own triangle of numbers, formed according to whatever rule you want to use. Does it have a practical value like counting something?

Level 2: Other patterns

Problem

What further patterns can your class see in Pascal's triangle?

Problem steps

Step 1

One obvious pattern is that the numbers increase and then decrease along the rows. In addition, they are symmetrical along the row.

Now we can look at some diagonals. The zeroth diagonal, going from the top 1 down to the left, isn't very interesting because it just consists of 1s. Hopefully the diagonal from the 1 at the end of the first row down to the left has more going for it. This diagonal consists of the numbers 1, 2, 3, 4, 5, 6 ... It is not much more interesting. This diagonal does intersect the nth row at n, though.

Ever optimistic, we move on to the next diagonal from the 1 in the second row down to the left. Here the numbers are 1, 3, 6, 10, 15... Have students seen these numbers before? Can we get a general term for them? From Level 3 of the *Tower of Hanoi* activity (Chapter 2), we can see that the member of this diagonal in the nth row is $\frac{n(n-1)}{2}$. These numbers are often referred to as the *triangular numbers*.

The next diagonal is 1, 4, 10, 20, 35 ... These turn out to be the sum of the triangular numbers up to that row. The reason for this can be found in Level 1. The member of this diagonal can be guessed to be $\frac{n(n-1)(n-2)}{(3 \times 2 \times 1)}$.

This suggests that the next diagonal, when it meets the nth row, produces the number $\frac{n(n-1)(n-2)(n-3)}{(4 \times 3 \times 2 \times 1)}$. Can students check this for a few values of n to see if it fits?

Step 2

Which leads us to ask: what is the formula for the general term in Pascal's triangle? What does your class think? It is starting to look like:

$$\frac{n(n-1)(n-2) \ldots (n-r+1)}{(r \times r-1 \times \ldots \times 3 \times 2 \times 1)}$$

Here the n tells us what row the number is in and the r tells us which diagonal it is on. But this is exactly the same as the counting that we did in Level 1, Step 5.

We can simplify the formula if we introduce the term $n!$, pronounced 'n factorial'. We let $n! = n(n-1)(n-2) \ldots \times 3 \times 2 \times 1$. So the general term in Pascal's triangle is $\frac{n!}{r!(n-r)!}$.

This is generally called a *binomial coefficient* and is denoted by ^nC_r as we have already seen. It is sometimes written as $\binom{n}{r}$.

Step 3

Now it is time to do a little algebra.

$$^nC_r + {}^nC_{r-1} = \frac{n!}{r!(n-r)!} + \frac{n!}{(r-1)!(n-r+1)!}$$

$$= \frac{n!}{(r-1)!(n-r)!} \left\{ \frac{1}{r} + \frac{1}{n-r+1} \right\}$$

$$= \frac{n!}{(r-1)!(n-r)!} \left\{ \frac{n-r+1+r}{r(n-r+1)} \right\}$$

So far this hasn't proved anything, though it does suggest that any entry in a row is the sum of the two entries above it in the previous row.

However, we can use mathematical induction to show that the entries of Pascal's triangle are indeed the binomial coefficients. We leave this for your more mathematically able students to try.

Step 4

Some of your students might add the numbers in any row together and find that they all add to powers of two. Can they prove that the sum of the numbers in the nth row add to 2^n?

This requires looking at Pascal's triangle from another direction.

We won't ask what $(x + 1)^0$ or $(x + 1)^1$ are, but will ask students to expand $(x + 1)^2$ and $(x + 1)^3$.

$$(x + 1)^0 = 1$$
$$(x + 1)^1 = x + 1$$
$$(x + 1)^2 = x^2 + 2x + 1$$
$$(x + 1)^3 = x^3 + 3x^2 + 3x + 1$$

Do the coefficients of the various powers of x remind them of anything? Get your class to predict what the expansions of $(x + 1)^4$ and $(x + 1)^5$ are.

So if $(x + 1)^n$ has coefficients from the nth row of Pascal's triangle, then put $x = 1$. Since $(1 + 1)^n = 2^n$, then the sum of the coefficients is 2^n. That means that the sum of the nth row of Pascal's triangle is 2^n.

Where to from here?

- Can you prove that the coefficients of $(x + 1)^n$ are the values in the nth row of Pascal's triangle? What else do you think is worth pursuing?
- This is a good time to give your students some counting problems that can be done using the binomial coefficients.

Level 3: Fibonacci's sequence

Problem

What is the Fibonacci sequence and where is it hidden in Pascal's triangle?

Problem steps

Step 1

There is a good chance that someone in the class will have heard of the Fibonacci sequence. Let them tell the class what they know, then let other students make contributions.

The basic thing that students need to know here is that the sequence starts off with 1, 1 and that to get the next number in the sequence you add the previous two numbers (see Figure 3.6). This will be the key aspect for the rest of this level.

Figure 3.6: The Fibonacci sequence and how it is constructed

1	1	2	3	5	8	13	21	34	...
	1+1	1+2	2+5	3+5	5+8	8+13	13+21	...

Spend some time playing with the sequence to make sure students gain some familiarity with it. For example, what are the 10th, 11th and 12th terms of the sequence, where are the even numbers, and so on. (The 10th term is 55, the 11th term is 89, the 12th term is 144, and the even Fibonacci numbers only occur in the 3rd, 6th, 9th, 12th etc. terms.)

But where do these numbers hide in Pascal's triangle? Think diagonals, think patterns.

Step 2

We know that the powers of 2 come from adding the numbers in a row of the triangle. Fibonacci numbers come from adding numbers in a diagonal, but it is not an obvious diagonal. Let your students work on possible diagonals to see if they can stumble across the diagonals we want.

The answer is in Figure 3.7.

Figure 3.7: Adding numbers on certain diagonals in Pascal's triangle

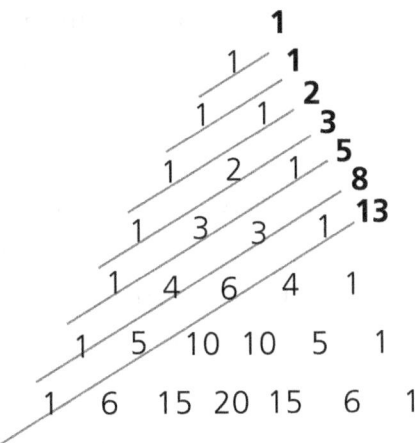

Get them to continue the pattern of Figure 3.7 to see that the sum of the numbers in these strange diagonals always gives a Fibonacci number.

This leads to the question: Why is it so? What suggestions can your students offer? How good are these ideas? Can they be made into a proof?

Step 3

Figure 3.8: A step towards why we get the Fibonacci numbers in Figure 3.7

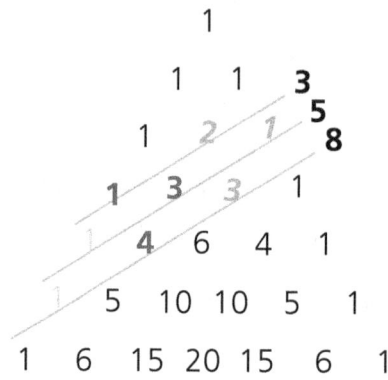

What we need to show is that the numbers in the diagonal whose sum is 3 combine in some systematic way with the diagonal whose sum is 5 to get all of the numbers in the diagonal whose sum is 8.

By the common property of the Pascal triangle, the italicised numbers 1 and 2 add to the italicised number 3. So we have used up the italicised numbers from the 3 and 5 diagonals. In the same way, the bold numbers 1 and 3 add up to the bold number 4. The only number in the 3 and 5 diagonals that hasn't been used so far is the underlined 1. We can just pass this down to the underlined 1 in the 8 diagonal. All of the diagonal numbers in the 8 diagonal can be found by adding numbers from the 3 and 5 diagonals.

Step 4

Repeat this argument with the 5 and 8 diagonals to get the 13 diagonal, then keep going.

Every time, the numbers from two consecutive strange diagonals will add together to get the numbers in the next diagonal. That is just what Fibonacci numbers do.

Making a general proof of this is hard, because it means keeping track of binomial coefficients, but some of your more able students may be able to put a proof together.

Where to from here?

- Can students find a formula for the Fibonacci numbers? (Try online.) Surprisingly, the formula contains $\sqrt{5}$.
- What do the students know about Fibonacci, what has this got to do with rabbits and how did his sequence of numbers come about?

Level 4: Getting in a team

Problem

Everything else being equal, what are the chances of your being chosen in a team of eight out of a squad of ten players?

Problem steps

Step 1

This requires some discussion to find the best method of attack. It might help to discuss a problem with smaller numbers before tackling this one.

The number of teams that can be chosen is $^{10}C_8 = 45$.

The number of teams that include you is the same as the number of teams of seven that can be chosen from nine players. This is $^9C_7 = 36$.

So the probability of your being chosen is $\frac{36}{45} = \frac{4}{5}$ or 0.8.

Is the fact that this is the size of the team divided by the number of possible players an accident?

Step 2

What if the captain (who isn't you) has to be in every team chosen? Are your chances of getting in a team better or worse than they were before? What is the exact probability of your getting in a team?

Discuss whether the chances of getting in the team are better or worse than in Step 1. Do this without calculating the exact probability involved. Who thinks it's better? Who thinks it's worse? Why? Listen to their reasons. Can they reach a conclusion without finding the exact probability involved?

Put students into groups to find an exact value and report back.

If the captain is in all teams then there is a choice of seven players from nine for the number of possible teams ($^9C_7 = 36$). The number of teams that include you is six players out of eight ($^8C_6 = 28$).

So the probability is $\frac{28}{36} = \frac{7}{9}$ or 0.77. The chances have *decreased* considerably. Was this predicted?

Once again, this looks like the size of the team divided by the number of players. Is there a pattern here?

Step 3

What is the chance of being chosen in a team with n players if there are $n + 2$ people to choose from? This is

$$\frac{^{n+1}C_{n-1}}{^{n+2}C_n} = \frac{n}{(n+2)}$$

Students should now see that as n grows, $\frac{n}{(n+2)}$ increases. In fact it is worth noting that as n gets bigger and bigger, $\frac{n}{(n+2)}$ gets closer and closer to 1. Is this what your students might have intuitively thought?

Step 4

If there are 11 players for a team of 8, will the probability of a given person getting in the team be $\frac{8}{11}$?

Discuss this with the class and then let the groups work out the answer—which, yes, is

$$\frac{^{10}C_7}{^{11}C_8} = \frac{8}{11}$$

Step 5

So does this work for n possible players for a team of size m with $m < n$? Is a given player's chance of getting in the team $\frac{m}{n}$?

Again, discuss this with the class as a whole and then set groups to work out the answer.

This is

$$\frac{^{n-1}C_{m-1}}{^{n}C_{m}} = \frac{\left(\frac{(n-1)!}{(m-1)!(n-m)!}\right)}{\left(\frac{n!}{m!(n-m)!}\right)}$$

This simplifies to $\frac{m}{n}$. This happens no matter what n and m are.

Where to from here?

- Is it clear why a given player's chances of getting in the team are $\frac{m}{n}$? What has this got to do with binomial coefficients?
- Get students to make up their own counting problems.
- 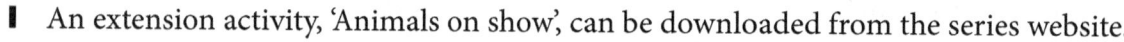 An extension activity, 'Animals on show', can be downloaded from the series website.

CHAPTER 9:
MONTY HALL'S PROBLEM

Initial problem

You are on a TV game show that hides a car behind one of three doors and a goat behind the other two. You are allowed to choose one door behind which you hope the car is waiting. The TV host, Monty Hall, then opens another door and there sits a goat. He asks you if you want to change your original choice of door.

Are you better off choosing the other door or staying with your original choice?

Background information
The Monty Hall problem has become well known because it is simple to state, hard to believe the solution, but relatively easy to explain the answer. Monty Hall was a real person—at various times an MC, producer, actor, singer and broadcaster. The problem is based on his TV show *Let's Make a Deal*, which originally aired in the 1960s. The problem has also been mentioned in the TV show *Numb3rs* and the book *The Curious Case of the Dog in the Night-Time*, as well as being subjected to scrutiny in *Mythbusters*. Its solution by columnist Marilyn vos Savant in 1990 caused considerable controversy, as even important mathematicians such as Paul Erdős did not initially believe the suggested best strategy.

»

All of the problems in this activity are based on problems whose solutions are not intuitively obvious. However, most of these solutions can be understood by most students and the exercise of studying them should increase their understanding of the subject.

There are many formulations of *Monty Hall's problem*, and Level 1 presents one of the most straightforward. Year 7 students should understand the Level 1 problem and see the solution, though many may have trouble doing the last few steps.

Level 2 presents a variant called *Bertrand's box paradox*, which goes back to 1889. Students who understood *Monty Hall's problem* can get started on Level 2, but a lot of Year 7 and Year 8 students will have difficulties with this problem.

Level 3, *Heads?*, uses the binomial coefficients from the *Pascal's triangle* activity to find the probability of getting a sequence of heads in some coin tossing. Level 3 is accessible for Year 9 students, but many Year 8 students should see how to do the first few steps, especially if they start off looking at fewer than six flips of the coin.

Level 4 *Two born every day* addresses the old problem of finding the probability of two people at a gathering having the same birthday. Level 4 is for Year 9 students.

Table 3.4: Australian Curriculum content descriptions for the *Monty Hall's problem* activity

Activity level	Problem	Content descriptions
1	Monty Hall's problem	*Year 7* Construct sample spaces for single-step experiments with equally likely outcomes (ACMSP167) Assign probabilities to the outcomes of events and determine probabilities for events (ACMSP168)
2	Bertrand's box paradox	*Year 7* ACMSP167 (see above) ACMSP168 (see above) *Year 8* Represent events in two-way tables and Venn diagrams and solve related problems (ACMSP292)
3	Heads?	*Year 8* Identify complementary events and use the sum of probabilities to solve problems (ACMSP204) *Year 9* List all outcomes for two-step chance experiments, both with and without replacement using tree diagrams or arrays. Assign probabilities to outcomes and determine probabilities for events (ACMSP225)
4	Two born every day	*Year 9* ACMSP225 (see above)

Big ideas
» Probability needs to be taken carefully
» Simulation as a useful tool
» Analysis of non-intuitive situations

Problem aims
» To put probability knowledge in interesting situations
» To see the value of simulation
» To introduce students to well-known statistical problems

Key concepts
» Simulation
» Probability in terms of binomial coefficients
» Sum of probabilities is 1

Possible heuristics/strategies
» Try simpler cases
» Look for patterns
» Guess and check (trial and error)

Resources
» Two goats and a car (or is this too much to expect?)

Special notes
Without loss of generality: This is a phrase used before a proof where restrictions seem to be being made but the restrictions are chosen so that they still cover all possibilities.

Paradox: This is a statement where there seems to be a contradiction within or outside the statement being considered.

Level 1: Monty Hall's problem

Problem

You are on a TV game show that hides a car behind one of three doors and a goat behind the other two. You are allowed to choose one door behind which you hope the car is waiting. The TV host, Monty Hall, then opens another door and there sits a goat. He asks you if you want to change your original choice of door.

Are you better off choosing the other door or staying with your original choice?

Problem steps

Step 1

This problem requires considerable thought. First, make sure all students understand the rules of the game. Then ask them whether they would change or not. They will need to justify their choice.

At first the answer seems obvious. There are only two choices open to you, the contestant, so it looks as if there is a 50-50 chance of success if you change and thus it doesn't matter whether you change or not.

It is likely that most students will say this. Take a vote on what you, as a contestant, should do.

Step 2

It is worth getting some evidence for a complete answer, so simulate the situation and see what data is produced.

The best way to do this is online; there are a number of simulation applets and animations that can be found via the series website.

Do as many trials as you have time for. If your students have access to a number of computers, then put them into groups and have each group do 20 trials. Pool the data and record the combined data on the board. Alternatively, if you have an interactive whiteboard and/or computer access is limited, you could do the trials for the whole class.

What does the data suggest is the best strategy?

Now take another vote on what your students, as contestants, should do. Has there been any change in the vote?

Surprisingly, changing the door appears to enable you to win the car more often. How much more often does your data suggest? Can this changing be quantified?

Step 3

In the light of the simulations, re-discuss the problem from a theoretical point of view. Can anyone give an argument that gives a probability of winning the car that approaches that of your simulation data?

We try to show what is going on in Figure 3.9, which can also be downloaded from the series website. Put a large version of this where all the students can see it and talk them through the explanation. Ask them if this explains why it is better to change doors. Give them a chance to work their way through the explanation.

In the initial situation, the three doors all look the same. Then the contestant makes her first choice. There are three possibilities, each shown by the bold line around the door chosen. Without loss of generality, we assume that the car is behind the door on the left. It should be clear that the chance of choosing the car is $\frac{1}{3}$.

If the contestant chooses the left door, Monty Hall can open either of the other doors to show the goat. This is shown using a large cross. (We only show one of these cases; they are equivalent for our purposes.) If the contestant chooses either of the other doors, Monty Hall is forced to open the door with the cross over it.

Figure 3.9: An explanation of the reason for change

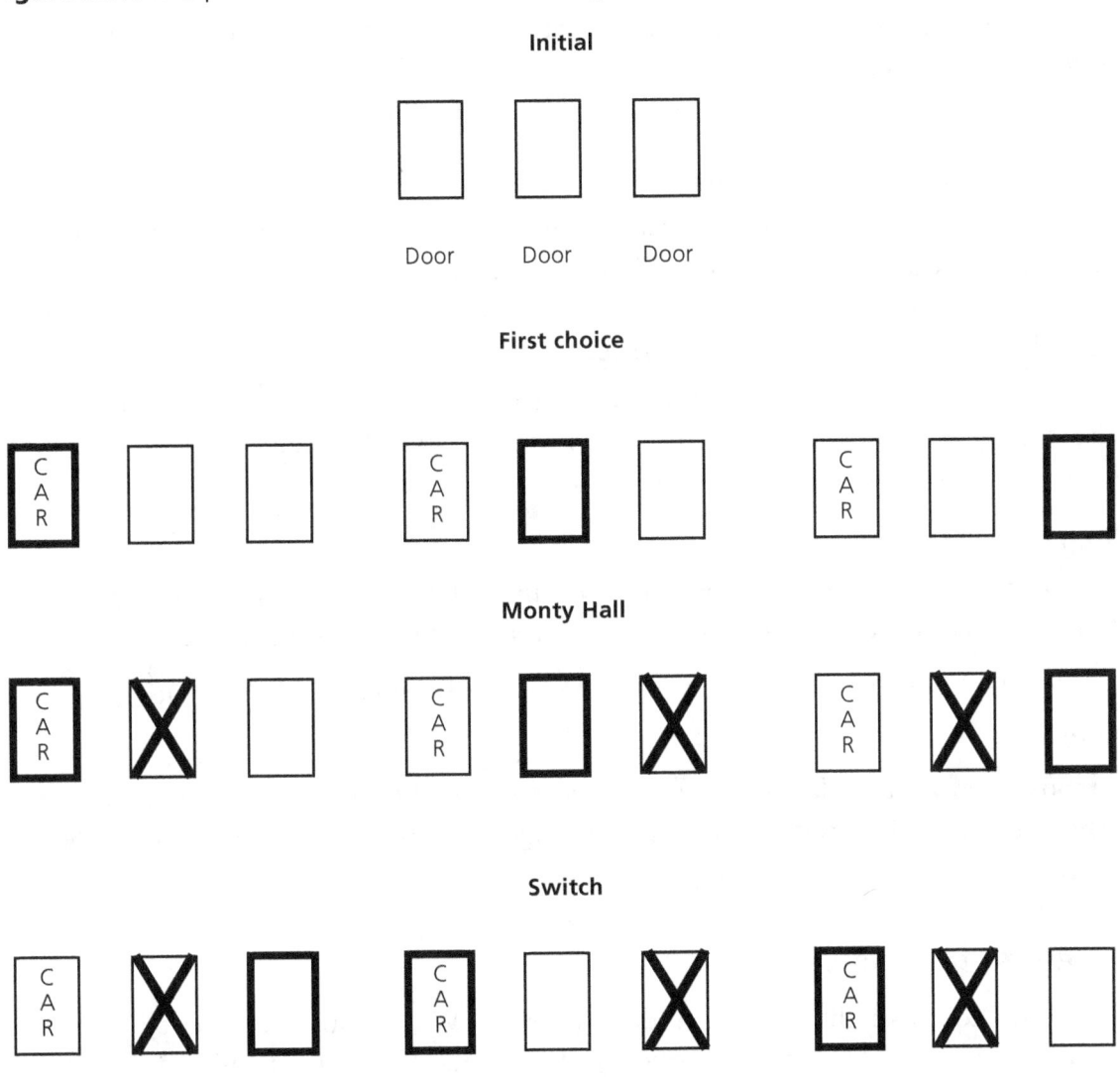

If the contestant changes her door choice at this point, she gets the car in two of the situations and loses the car in the other. So her chance of getting the car is $\frac{2}{3}$. It is clearly better for her to change the door she wants to open.

To consider it without a diagram: the contestant initially chooses the car door with a probability of $\frac{1}{3}$. If she waits until Monty Hall shows the goat behind one of the doors there are now three situations, with two of them having the car behind the doors the contestant didn't choose. If she changes doors she has a probability of $\frac{2}{3}$ of getting the car.

The point is that by showing the goat, Monty changes the initial conditions of the contest from one where the chances of getting the car are $\frac{1}{3}$ to one where the chances are $\frac{2}{3}$.

This game is clearly counter-intuitive, since most people seeing it for the first time conclude that it doesn't matter if a contestant changes or not.

Step 4

In Figure 3.9 we put the car behind the left door and said that this was *without loss of generality*. In other words, it didn't matter where we put the car, the result would be the same.

Get the students to draw their own diagram to show that it really doesn't matter which door is chosen first.

Step 5

We can now consider a variation.

This time, Monty Hall hides a car behind one of *four* doors and a goat behind the other three. You choose one door; Monty Hall then opens two other doors to reveal a goat behind each. What should you do now? Are you better off choosing another door or should you stay with your original choice?

Once again, you should change; this changes your chance of success from $\frac{1}{4}$ initially to $\frac{3}{4}$ after Monty Hall's intervention. A diagram such as Figure 3.9 should help your students to see this if they struggle.

Step 6

We can go even further—now there are *one hundred* doors. Once you choose your door, Monty opens 98 doors to reveal 98 goats. What's your best chance of success?

In this instance, changing alters your success rate from $\frac{1}{100}$ to a massive $\frac{99}{100}$.

Step 7

Suppose there are four doors; you choose one, then Monty opens one door to reveal a goat, leaving two closed. How do your chances change this time if you choose a new door?

This time you increase your chances from $\frac{1}{4}$ to $\frac{3}{8}$.

Where to from here?

- Discuss the possibilities for a five-door TV show. What should the rules be? How many doors should Monty Hall open to show goats? How does opening different numbers of doors change the probability of success?

- Suppose that you are a darts player playing in a championship. You are given three identical envelopes, each of which has the name of one opponent that you might play next. One of these is a player you know you can beat, so you'd like to play him. After you choose an envelope, the director of the championship removes one of the remaining envelopes. Does changing your choice of envelope alter your chances of playing the poorer player?

Level 2: Bertrand's box paradox

Problem

Three cards of identical size are put in a hat. One card has both faces white, another has both faces black, and the third card has one white and one black face.

A card is drawn at random from the hat and placed on the table. The side facing up is black. What is the probability that the other side is black?

Problem steps

Step 1

Bertrand's box paradox (named after 19th-century French mathematician Joseph Bertrand) is conceptually similar to *Monty Hall's problem*. We give this in the form of a card problem as it may be easier to see; it is certainly easier to simulate.

Again, the first response for most students is that the probability that the other side is black is $\frac{1}{2}$. This is based on the fact that there are *two* cards where the answer is black—the card with both sides black and the card with one side of each colour. Since we seem to be choosing one from two, the probability has to be $\frac{1}{2}$.

Get your students to guess the answer first. (You might take a vote again to see how good their intuition is.) Then make the cards as described and simulate the card drawing to see what experimental probability they come up with.

Step 2

What answer do your students get from the simulation? Can they prove this theoretically?

Label the faces of the white card W1 and W2; the faces of the black card B1 and B2; and the faces of the mixed card W3 and B3. The black face that is shown on the table may be B1, B2 or B3.

Now refer to Table 3.4, which uses tick marks to show what the other face might be.

Table 3.4: Possible outcomes of *Bertrand's box* problem

Face showing	Possible hidden faces					
	B1	B2	B3	W1	W2	W3
B1	✗	✓	✗	✗	✗	✗
B2	✓	✗	✗	✗	✗	✗
B3	✗	✗	✗	✗	✗	✓

There are three possible outcomes:
1. B1 is the black face we can see on the table and B2 is the other face.
2. B2 is the black face we can see on the table and B1 is the other face.
3. B3 is the black face we can see on the table and W3 is the other face.

There are three possible situations. In two of these the hidden face is black, so the probability that the hidden face is black is $\frac{2}{3}$.

Step 3

Bertrand's original box paradox was phrased this way:

Three boxes each have two drawers. In the first box, the drawers each contain a gold coin; in the second box, each drawer contains a silver coin; and in the last box one drawer has a gold coin and one a silver coin. A box is chosen at random and a random drawer is opened. A gold coin is found inside. What is the probability that the other coin in that box is silver?

Let your students discuss this. From the discussion in Step 2, it should be clear that the probability is $\frac{1}{3}$.

Step 4

What happens in Bertrand's box paradox if there are *three* drawers in each of three boxes?

Suppose that in the first box each drawer has a gold coin; in the second box each drawer has three silver coins; and in the third box, there are two drawers with gold coins and one with silver. Choose a box at random. Randomly choosing first one drawer and then a second from this box reveals two gold coins. What is the probability that the other drawer contains a silver coin?

What do your students think is the answer? By now they shouldn't assume that the answer is $\frac{1}{2}$. Do they have good reasons to support their arguments? Are there any counter-arguments?

Step 5

The box chosen must either be the first box or the third box. There are six ways of choosing the two gold coins from the first box, so there are six ways that the final choice would be gold. There is only one way of choosing two gold coins from the third box, so there is only one way that the third choice is a silver coin.

The probability that the third drawer contains a silver coin is therefore $\frac{1}{7}$.

Where to from here?

- The possibilities for producing more box problems are endless. Let your students make up and solve their own problem on this theme.

- The series website has links for some similar problems—the 'Two envelopes' problem, the 'Sleeping beauty' problem and the 'Three prisoners' problem. Can your students work through these?

- Bertrand has another paradox, which is much more perplexing: Draw the circumcircle (see Chapter 3) of an equilateral triangle. Choose a chord at random. What is the probability that the chord is longer than a side of the triangle? It turns out that there are at least three ways to do this question, and they all end up with a different answer.

Level 3: Heads?

Problem

What is the probability of getting *exactly* two heads in six flips of a coin?

Problem steps

It might be good to start off with fewer than six flips to help students see what is going on here.

Step 1

The simplest way to do this is to list all the possible outcomes and choose the ones that have just two heads. The list of outcomes might start like this:

HHHHHH

HHHHHT, HHHHTH, HHHTHH, HHTHHH, HTHHHH, THHHHH

HHHHTT, HHHTHT, ...

This is a bit tedious, as there are 64 possibilities. (How can students determine this?) However, once you have them all you can choose the ones we're interested in. There are 15 of these, so the answer is $\frac{15}{64}$, or about a quarter.

Get the class to discuss how Pascal's triangle can be used here. Note that 64 is a power of 2 (in fact 2^6), which makes it the sum of all of the binomial coefficients in the sixth row. And 15 is the number of ways of choosing 2 from 6, or $^6C_2$.

Step 2

Given the problems that came up in the Monty Hall problem and paradoxes, it is worth checking that all of the possible probabilities in six tosses of a coin add up to 1. So does

  the probability of getting exactly no heads in six flips of a coin
+ the probability of getting exactly one head in six flips of a coin
+ the probability of getting exactly two heads in six flips of a coin
+ the probability of getting exactly three heads in six flips of a coin
+ the probability of getting exactly four heads in six flips of a coin
+ the probability of getting exactly five heads in six flips of a coin
+ the probability of getting exactly six heads in six flips of a coin
= all sum to 1?

Students should find that the probabilities of:

- getting exactly no heads in six flips of a coin = $\frac{1}{64}$
- getting exactly one head in six flips of a coin = $\frac{6}{64}$
- getting exactly two heads in six flips of a coin = $\frac{15}{64}$
- getting exactly three heads in six flips of a coin = $\frac{20}{64}$
- getting exactly four heads in six flips of a coin = $\frac{15}{64}$
- getting exactly five heads in six flips of a coin = $\frac{6}{64}$
- getting exactly six heads in six flips of a coin = $\frac{1}{64}$.

These numbers come in the same way that they did in Step 1. Furthermore, the sum is definitely 1, since 1 + 6 + 15 + 20 + 15 + 6 + 1 = 64.

How could we have known this without doing any addition? (By referring to Pascal's triangle; this is the sum of the rows of the triangle.)

Step 3

What is the probability of getting *exactly* two heads in sixteen flips of a coin?

Your students *definitely* don't want to do this by listing all possibilities. For a start, there are $2^{16} = 8192$ possibilities to list! So we need to find some other way to tackle this problem.

First, how do we know how many possibilities there are? Get the students to try to count these. If they can't see it immediately, assume that you flip the coins once, twice and so on and draw up a table to see the pattern. They should come up with 2^n ways for flipping n times.

Another way to see this is to do it directly. Write the possibilities in a bracket format as (X, X, X, … X). The X before the first comma tells you what happened after the first coin was tossed, and the same thing continues through the bracket for the n Xs. Each X can be a head or a tail, so each X can have two values. Altogether there are n Xs, so there must be $2 \times 2 \times \ldots 2$ (with n 2s) $= 2^n$.

Secondly, once you know how many possible outcomes there are, how do you pick out the number that have two heads? This is just choosing two things out of 6 (in Step 1), or 16 (in this problem) or n (in general). And that is $^6C_2$, or $^{16}C_2$, or $^nC_2$.

This means that the probability of getting two heads in 16 tosses of the coin is $\frac{^{16}C_2}{2^{16}}$. This is also equal to the number of ways of choosing two things from 16 ($^{16}C_2$) multiplied by the probability of getting two heads $(\frac{1}{2})^2$ multiplied by the probability of getting fourteen tails $(\frac{1}{2})^{14}$.

Step 4

Once again, check to see that all of the possible probabilities in 16 tosses add up to 1. So, do:

   the probability of getting no heads in 16 tosses of the coin
+ the probability of getting one head in 16 tosses of the coin
+ the probability of getting two heads in 16 tosses of the coin
+ the probability of getting three heads in 16 tosses of the coin
+ the probability of getting four heads in 16 tosses of the coin
+ the probability of getting five heads in 16 tosses of the coin
+ the probability of getting six heads in 16 tosses of the coin
+ the probability of getting seven heads in 16 tosses of the coin
+ the probability of getting eight heads in 16 tosses of the coin
+ the probability of getting nine heads in 16 tosses of the coin
+ the probability of getting ten heads in 16 tosses of the coin
+ the probability of getting eleven heads in 16 tosses of the coin
+ the probability of getting twelve heads in 16 tosses of the coin

+ the probability of getting thirteen heads in 16 tosses of the coin
+ the probability of getting fourteen heads in 16 tosses of the coin
+ the probability of getting fifteen heads in 16 tosses of the coin
+ the probability of getting sixteen heads in 16 tosses of the coin
= all add to one?

This isn't as hard as it looks, because:

$$^{16}C_0 + {}^{16}C_1 + {}^{16}C_2 + {}^{16}C_3 + {}^{16}C_4 + {}^{16}C_5 + {}^{16}C_6 + {}^{16}C_7 + {}^{16}C_8$$
$$+ {}^{16}C_9 + {}^{16}C_{10} + {}^{16}C_{11} + {}^{16}C_{12} + {}^{16}C_{13} + {}^{16}C_{14} + {}^{16}C_{15} + {}^{16}C_{16} = 2^{16}$$

Why is that? (Refer back, once again, to Pascal's triangle.)

Step 5

What is the probability of getting *exactly* one 6 in five rolls of a die?

Discuss this in class, then get students to work in groups to experiment and answer.

The answer here is not $\frac{^5C_1}{6^5}$ as they might expect. The difference here is that dice provide more than two options, and this has to be taken into account. So a 6 has a chance of $\frac{1}{6}$ and all the others together have a chance of $\frac{5}{6}$.

If there are two 6s in the first two rolls, that gives you a chance of $(\frac{1}{6})^2 \times (\frac{5}{6})^3$. But it is the same in the first and third rolls, and all of the other rolls you can have. There are $^5C_2$ ways to choose two positions, so the neatest way to express the answer is as $^5C_2 \times (\frac{1}{6})^2 \times (\frac{5}{6})^3$.

Step 6

It would be good to make sure that all of the possible probabilities of rolling five dice add up to 1.

So does $^5C_0 \times (\frac{1}{6})^0 \times (\frac{5}{6})^5 + {}^5C_1 \times (\frac{1}{6})^1 \times (\frac{5}{6})^4 + \ldots + {}^5C_5 \times (\frac{1}{6})^5 \times (\frac{5}{6})^0 = 1$?

Now, think about expanding $(1 + 5)^5$. From the binomial expansions in Level 2, this is $^5C_0 \times 1^0 \times 5^5 + {}^5C_1 \times 1^1 \times 5^4 + \ldots + {}^5C_5 \times 1^5 \times 5^0$. But $(1 + 5)^5$ is clearly 6^5. And that is how we get that 1.

Where to from here?

▍ What is the probability of getting exactly *h* heads in *n* flips of a coin?

▍ What is the probability of getting exactly *s* 6s in *r* rolls of a die?

▍ If we are looking for the probability of getting exactly *s* of something with probability *p* in *t* trials, then it is just $^tC_s \times p^s \times (1-p)^{s-t}$. This leads to what is known as the *binomial distribution*.

Level 4: Two born every day

Problem

What is the probability that two of the people in your classroom right now have their birthdays on the same day of the year?

Problem steps

We should be clear about this. They don't have to have been born on the *same* day of the *same* year. We only want them born on the same day of *some* year.

(By 'year' we mean a normal year. For this activity, assume that Leaplings were born on 1 March.)

Step 1

Discuss how the class might go about getting a solution. Are there two people in the class today (including you) that have their birthday on the same day? Do they think that that is unusual?

Try some extreme cases. For example, is it possible that in some groups of people you can be *absolutely sure* that there are two people with the same birthdays? Yes, if the group had 365 or more people in it. This is an example of the 'pigeonhole principle'; if you have n pigeonholes and more than n pigeons, at least two pigeons must be in one hole.)

Simplify again. What is the probability that two people *chosen at random* have their birthdays on the same day?

Do they need another simplification? What is the probability of getting two heads *or* two tails if you toss two coins? (It is $\frac{1}{2}$; confirm this by looking at all the ways two coins can be thrown.) Now get gradually more complicated. In the game Rock-Paper-Scissors, what is the chance of two players choosing the same thing? ($\frac{1}{3}$.) What is the probability of rolling the same number with two dice? ($\frac{1}{6}$.)

What is the probability, then, of two people having the same birthday? It is $\frac{1}{365}$, but how can they be sure? What does the pattern we have been building up suggest? (The first person can be born on any day; the second has a $\frac{1}{365}$ chance of being born on that same day.)

Can we generalise to any number of people? How?

Step 2

Now we can build things up again. If we have three coins and want at least two heads or two tails, there is no problem. The pigeonhole principle shows that there is a probability of 1.

How about Rock-Paper-Scissors with three players? What is the probability of two players choosing two of the same? If the class is listing all possibilities, they should find 21. So the probability we're looking for is $\frac{21}{27} = \frac{7}{9}$.

What about rolling two of the same numbers if we roll three dice? ($\frac{4}{9}$, though this is starting to get tough if they use the list-of-possibilities method.) It is even tougher to work out accurately how many ways there are of getting two of the same with four, five or six dice. ($\frac{2}{3}$, $\frac{49}{54}$ and $\frac{643}{648}$.)

Why are these numbers getting closer and closer to 1? Because there are fewer and fewer ways to avoid getting a match as you increase the number of dice up to seven.

How about three people having the same birthday? That is getting hard and for four people it is even harder. There has to be a better way.

Step 3

The direct approach is getting harder and harder, so we could try the indirect approach.

How else can your students think of looking at this problem? Can they test their ideas on what is the probability of three players all getting *different* things in Rock-Paper-Scissors?

This can be counted as before, but we can also calculate it directly with a little thought. One of the players has made a gesture—let's call it X. So the second player has two ways to *not* get X, and she can do this with a probability of $\frac{2}{3}$. The final player can miss out on X with a probability of $\frac{1}{3}$. So the probability of making all three different gestures is $\frac{2}{3} \times \frac{1}{3} = \frac{2}{9}$. This means that the probability of getting two the same is $1 - \frac{2}{9} = \frac{7}{9}$.

Now try using this method on dice. The students should keep a careful record of the fractions they are multiplying together so that they can generalise.

For 'no two numbers the same' with four, five and six dice it is $\frac{5}{6} \times \frac{4}{6} \times \frac{3}{6} = \frac{10}{36}$, $\frac{5}{6} \times \frac{4}{6} \times \frac{3}{6} \times \frac{2}{6} = \frac{10}{108}$ and $\frac{5}{6} \times \frac{4}{6} \times \frac{3}{6} \times \frac{2}{6} \times \frac{1}{6} = \frac{10}{648}$, respectively.

Step 4

What is the probability of two people having the same birthday if there are three, four, five or six people in the room? It is time to forget fractions and get decimal answers—they are easier to calculate.

Now consider a classroom with 23 people in it. (You can find a demonstration of this on the series website.)

The probability that there are no shared birthdays in the group

$$= \frac{364}{365} \times \frac{363}{365} \times \frac{362}{365} \times \frac{361}{365} \times \ldots \times \frac{344}{365} \times \frac{343}{365}$$
$$= 0.4927\ldots$$
$$= 49.3\% \text{ (rounded)}$$

Therefore the probability that at least two people share the same birthday is 100% − 49.3% = 50.7%.

Counter-intuitively, it is actually more likely than not than in a group of 23 or more people, two or more of them will share a birthday!

It is possible to write the general form of this for p people, using a spreadsheet to draw the graph of the function as p gets larger.

Where to from here?

- With current technology, the probability that two people may be found to have the same DNA is about $\frac{1}{7000}$. What does that mean? What is the probability that two people seem to have the same DNA? What is the significance of this for DNA testing in court trials? What does this mean when we are told that everyone's DNA differs from everyone else's? Students could dig into this further as an online investigation.
- An extension activity, 'Born on a Tuesday', can be downloaded from the series website.

CREATIVE ACTIVITIES IN **MATHEMATICS**

Problem-based learning is a powerful alternative to drill-and-practice or skills-based learning, especially within mathematics.

The *Creative Activities in Mathematics* series provides a wealth of investigations and open-ended active learning activities, designed to engage students with mathematics and develop their problem-solving, collaboration and mathematical skills.

The three titles in the series provide a variety of class activities suitable for students from lower primary to middle secondary, along with teaching notes and staged lesson plans. Each activity is a whole-class investigation with open-ended answers that takes a particular scenario and develops it over multiple levels. This enables it to be used both at different year levels and with students of differing ability in the same class. All activities are firmly grounded in the Australian Curriculum: Mathematics.

Links to extra information, activities and student worksheets are available and easy to access online.

About the authors

Derek Holton is a mathematician and an Honorary Professor at the Melbourne Graduate School of Education.

Cath Pearn is a Senior Research Fellow in the ACER Institute and a lecturer in Mathematics Education at the University of Melbourne.

Duncan Symons is a Lecturer of Science and Mathematics Education at the University of Melbourne.

Charles Lovitt has directed several Australian national and state mathematics projects and is now a consultant and workshop presenter.

Amba Press | www.ambapress.com.au | hello@ambapress.com.au

www.ingramcontent.com/pod-product-compliance
Lightning Source LLC
Chambersburg PA
CBHW081102070526
44584CB00021B/3178